Analyzing
the Labor Force

Concepts, Measures, and Trends

PLENUM STUDIES IN WORK AND INDUSTRY

Series Editors:
Ivar Berg, *University of Pennsylvania, Philadelphia, Pennsylvania*
and Arne L. Kalleberg, *University of North Carolina, Chapel Hill, North Carolina*

WORK AND INDUSTRY
Structures, Markets, and Processes
Arne L. Kalleberg and Ivar Berg

Current Volumes in the Series:

ANALYZING THE LABOR FORCE
Concepts, Measures, and Trends
Clifford C. Clogg, Scott R. Eliason, and Kevin T. Leicht

EMPLOYMENT RELATIONS IN FRANCE
Evolution and Innovation
Alan Jenkins

NEGRO BUSINESS AND BUSINESS EDUCATION
Their Present and Prospective Development
Joseph A. Pierce
Introduction by John Sibley Butler

THE OPERATION OF INTERNAL LABOR MARKETS
Staffing Practices and Vacancy Chains
Lawrence T. Pinfield

SEGMENTED LABOR, FRACTURED POLITICS
Labor Politics in American Life
William Form

THE SOCIAL AND SPATIAL ECOLOGY OF WORK
The Case of a Survey Research Organization
Rita Gorawara-Bhat

SOURCEBOOK OF LABOR MARKETS
Evolving Structures and Processes
Edited by Ivar Berg and Arne L. Kalleberg

STRESS AND DISTRESS AMONG THE UNEMPLOYED
Hard Times and Vulnerable People
Clifford L. Broman, V. Lee Hamilton, and William S. Hoffman

WORLDS OF WORK
Building an International Sociology of Work
Edited by Daniel B. Cornfield and Randy Hodson

A Chronological Listing of Volumes in this series appears at the back of this volume.

A Continuation Order Plan is available for this series. A continuation order will bring delivery of each new volume immediately upon publication. Volumes are billed only upon actual shipment. For further information please contact the publisher.

Analyzing the Labor Force
Concepts, Measures, and Trends

Clifford C. Clogg
Late of Pennsylvania State University
University Park, Pennsylvania

Scott R. Eliason
University of Minnesota
Minneapolis, Minnesota

and

Kevin T. Leicht
University of Iowa
Iowa City, Iowa

Kluwer Academic / Plenum Publishers
New York · Boston · Dordrecht · London · Moscow

Library of Congress Cataloging-in-Publication Data

Clogg, Clifford C.
  Analyzing labor markets: concepts, measures, and trends/Clifford C. Clogg, Scott R. Eliason, and Kevin T. Leicht.
      p. cm. — (Plenum studies in work and industry)
  Includes bibliographical references and index
  ISBN 0-306-46536-1 — ISBN 0-306-46537-X (pbk.)
      1. Labor—United States. 2. Labor—United States—History. I. Eliason, Scott R. II. Leicht, Kevin T. III. Title. IV. Series.

HD8072.5 .C65 2001
331.11'0973—dc21

                                                                          00-067431

ISBN 0-306-46536-1 (Hardbound)
ISBN 0-306-46537-X (Paperback)

©2001 Kluwer Academic / Plenum Publishers, New York
233 Spring Street, New York, N.Y. 10013

http://www.wkap.nl/

10 9 8 7 6 5 4 3 2 1

A C.I.P. record for this book is available from the Library of Congress

Preface

With the passing of Clifford Collier Clogg at the age of 45 on May 7th 1995, the world lost a talented sociologist, demographer, and statistician all at once. In addition to being a considerable talent in each of these three disciplines, and perhaps more importantly, Cliff was the type of person who brought together diverse elements and scholars from all three. Cliff was also a consummate mentor, nurturing ideas and students and always striving to bring out the best in both. Perhaps nothing illustrates the stature, impact, and respect others held for Cliff more than the fact that never before—and never since—has an individual been honored at the time of his death with ceremonies from the national associations of all three of these disciplines.

The purpose of this book is to introduce to a broad constituency of social scientists and their students some of the basic ideas in the study of the labor force that Cliff and his colleagues had grappled with. At the time of Cliff's death, he was perhaps better known for his methodological contributions to sociology and demography than he was for his substantive contributions to the study of social stratification and the labor force. Our goal is to highlight Cliff's substantive contributions to sociology and demography by telling the cumulative story of his research and adding updated analysis that advances the story beyond the early 1980s to the mid-1990s.

THE EVOLUTION OF THE IDEA FOR THIS BOOK

Cliff's overall contribution to the study of labor force demography and social stratification needs to be evaluated in light of his entire scholarly career. Cliff's major research agenda could be summarized under three headings: (1)

the general analysis of demographic rates and statistical methods for adjusting rates; (2) introducing latent class analysis to the social sciences and demonstrating their utility for answering a variety of questions; and (3) substantive and policy relevant analyses of the labor force and the incorporation of labor force analysis into the study of social stratification.

As his career progressed, Cliff spent relatively more time addressing the methodological dimensions of his research agenda and relatively less time on the substantive implications of his work. We are not sure that this is a characterization of his work of which all those associated with Cliff would approve, but it is reflected in his accumulation of research products over the 1980s and early 1990s. Relatively few of his substantive analyses use data collected after 1983 and many analyses use data from the 1970s. Many of Cliff's colleagues at Penn State believed that the substantive implications of Cliff's work in labor force demography and the analysis of labor market opportunities were not getting the credit they deserved and were receding into the background among the social science community; some actually had conversations with Cliff about this. Because of the overall shock and lack of warning accompanying Cliff's sudden passing, these conversations had not advanced beyond the level of casual coffee-break discussions among workplace colleagues.

Of course, the gradual shift in Cliff's work was partly a byproduct of his caring and giving attitude toward the larger social science community. In addition to serving on numerous NSF and grant review panels and commenting on and improving his colleagues' research at Penn State and elsewhere, Cliff gave seminars and consulted on statistical and methodological issues with scholars and researchers all over the world. These activities would take him far from State College on a weekly basis. While (from all appearances) he enjoyed these activities, they did place serious constraints on his time and energies that probably prevented the further development of his substantive research agenda.

Cliff's death shocked the worldwide social science community in a way that few other events have. Partly because of the general outpouring of grief and the gathering of hundreds of people who were Cliff's friends and associates, and because of prior discussions about the relative neglect of Cliff's substantive contributions to the study of the labor force, we approached Eliot Werner and Arne Kalleberg at Plenum Press with the idea of producing a tribute to Cliff's substantive work. We knew that there were tributes planned in the *Journal of the American Statistical Association* and *Sociological Methodology*, but we knew of no individual or group who were preparing a substantive tribute.

There were two options available for producing a commemorative volume on Cliff's substantive work. We could assemble a standard tribute, soliciting contributions from current and former students and research colleagues most familiar with Cliff's work and asking them to evaluate the enduring contributions of specific dimensions of it. This alternative would have been fine, but it would look and feel like a standard festschrift that would appeal to those who knew Cliff but to few others.

Given Cliff's sudden passing and the state of limbo in which his research was left, we decided to pursue a different course. We decided to take a sampling of Cliff's substantive contributions to labor force demography and social stratification, scattered in different locations and written for different specialized audiences, and construct a cumulative narrative from them. This narrative would highlight the substantive contributions of Cliff's work that we hoped someone else would pick up on and continue; it would also allow us to take collective stock of where Cliff's substantive research program stood so we could make recommendations for continuing in the research tradition in which he worked. Eliot Werner and Arne Kalleberg graciously agreed to this idea.

Because much of Cliff's substantive work uses data from the late 1970s and early 1980s, we felt that a fitting tribute to him would require updated analyses. Our volume provides two updated original analyses that attempt to move the analysis of the labor force and labor force careers into the late 1990s. The work necessary to do this partly explains why it has taken this long to produce our book. The other reasons are more standard and mundane: we each had ongoing research agendas of our own and we had to make time to do this in what was an otherwise crowded academic calendar.

Our decision to do more than a standard tribute for Cliff has led to some non-standard conventions in this book of which readers should be aware. The most obvious of these is that Cliff is listed as an author of a book he did not write. We did this because we could think of no other way of honoring him for his contributions and crediting him for his work that we survey and add to here.

The other nonstandard convention is the use of "we" throughout, which refers to Clogg, Eliason, and Leicht as a collective group in the present even though we (Eliason and Leicht) were not collaborators on much of the original research. This seemed preferable to repeating the names of Clogg and his many coauthors throughout the text and it emphasizes our original desire for Cliff to write a cumulative treatise on his contributions to the study of the labor force. As fate would have it, we are doing this for him. We hope that you will view this work as a retrospective (perhaps spiritual) collaboration and that we have not overstepped our bounds. Obviously, not everyone familiar with Cliff's work would choose to summarize it as we have. Any faults you find in this presentation are due to us (Leicht and Eliason) and should not be attributed to Cliff.

In our presentation we make no presumption of detachment toward our colleague's work. Cliff was a close friend and colleague of many who will read this book and we wrote this book as a tribute to the person we knew and the considerable achievements he made during his short career. Instead, we have attempted to display the cumulative impact of that work in as favorable a light as possible and in a forum where a wider audience would be exposed to it. If people take this book off of their shelves and remember the man whose work we present, think of how to improve on it, and forget our rather modest attempts at summarizing and advancing it, then we will be more than happy with the results. If graduate students and others interested in social stratification and labor markets become interested in labor force issues as a result of this book, that would be better still.

A BRIEF NOTE OF THANKS

Although our names appear on the cover, we simply could not have brought this all together without the help and guidance of friends and colleagues. Teri Fritsma and Heather Wendt were instrumental in the thankless job of cleaning and compiling the data and text. Dennis Hogan provided invaluable feedback, especially at the early stage of the project. Beth Lyman patiently copyedited the final version of the text. We also wish to recognize the efforts of the many coauthors with whom Cliff worked on these pieces, especially James Shockey. To Judy Clogg, who helped us in countless ways and whose loss is far greater than ours, a very warm and heartfelt thanks; all of the proceeds from the sales of this book will go to Judy and her family.

We would also like to thank Arne Kalleberg and Eliot Werner for sticking with us through what turned out to be a much more involved and longer venture than anyone would have guessed at the beginning of our journey. And finally, we would like to thank our spouses, who put up with far more than our students or colleagues will ever know or would want to know.

Scott R. Eliason
Kevin T. Leicht

Contents

**Section IV:
Social Mobility, Socioeconomic Attainment and Labor Force Issues**

**Section V:
Recent Analyses of Labor Force Trends Using the Labor
Utilization Framework**

Section VI:
A Future Research Agenda

Chapter 9
Toward a More Complete Understanding of Labor-Markets

1

Introduction

Researchers, policymakers, and the broader public have concluded that the U.S. labor market has changed drastically in the last 20 years. Newspapers and magazines are crowded with articles asking basic questions about unemployment, job quality, loss of status and income, and the uncertainty of future labor markets. Our attention is riveted to this national conversation because unemployment and lost opportunity are intimately tied to individual life stories. The three life stories below demonstrate the importance of studying labor market activity in the United States and also illuminate the assets and liabilities of current thinking on these issues.

Our first story deals with Joe, an employee of U.S. Steel Corporation in Pittsburgh, Pennsylvania, in 1975. Joe makes steel in an open hearth furnace and has worked for U.S. Steel for 20 years. His father (recently retired) and grandfather worked for U.S. Steel as well. Joe's grandfather was involved in the formation of the United Steelworkers of America and could remember the struggles that organized labor went through to organize the steel industry. It is safe to say that Joe and U.S. Steel go way back. Connections with the company form a rich tradition in Joe's family life.

Joe's position with U.S. Steel is well paid, with wages around $18 an hour plus benefits (which effectively double his total compensation from U.S. Steel). Overtime is fairly routine, and Joe can easily count on working enough overtime every year to earn 20–40% more than his regular, full-time wage. His earnings are more than enough to afford a middle class lifestyle for Joe and his family: a suburban home, two cars, a small boat, and regular vacations.

Now fast-forward the scene to 1983. Joe is laid off at U.S. Steel and has

*Portions of this chapter were presented at the annual meetings of the Population Association of America, New Orleans, Louisiana, May 9–11, 1996.

1

difficulty finding another job. In part this is due to the general lack of employment opportunities of any kind in the Pittsburgh area. But Joe's inability to find a job also stems from two other factors, namely, his inability to find a job with the pay and benefits he had while working for U.S. Steel, and his unwillingness to uproot his family from the Pittsburgh area to find better work opportunities.

After nearly exhausting his unemployment compensation, Joe takes a job as a maintenance man for an apartment complex. His new job pays less than half of his former wage. Joe now works fewer hours and there are fewer opportunities for overtime. Joe sells his small boat and one of his cars. He also takes a job moonlighting as a security guard to make ends meet for his family.

Now consider our second life history. Mary is a secretary for a temporary help agency. The pay from the agency is fairly good ($9/hour), but the benefits package is limited. Mary's job does allow her to work flexible hours. She works on days when she wants to and passes on new assignments when she doesn't want to take them. Mary has a bachelor's degree in English, something that almost none of the other workers in the temporary help agency have, including her boss. When asked why she remains with the agency, Mary replies that the money is fairly good, her husband already has a good job with generous fringe benefits, and she cannot relocate to an area where the skills of her college degree would be better utilized. Besides, there is variety in her job because she works for different employers doing a wide range of tasks. Generally, Mary paints a picture that suggests that this is the best job she can get.

Our third life history may sound more familiar to you. Michael is someone who all of the major newspapers have had something to say about (not to mention politicians and academics). He lives in poverty. He goes from one dead-end minimum wage job to the next. The jobs rarely if ever last longer than a year, and Michael is as likely to quit jobs he doesn't like as he is to remain employed. The hours he works are extremely variable, but rarely full-time. It is not unusual for Michael to be unemployed for several months during the year, during which time he often engages in illegal activities to supplement his meager income from unemployment insurance ("running numbers," loan sharking, petty drug dealing, and so on). Michael has no recognizable job skills and is functionally illiterate. He stopped going to school in the eighth grade and officially dropped out at 16. The chances that Michael will ever "make it" (in conventional terms) seem very slim.

These are three of countless millions of biographies describing contact with the labor market in the United States. Each presents complications for researchers interested in understanding current and future labor market opportunities in the United States. Yet our current state of knowledge leaves us with difficulties in dealing with these complications. Consider the following:

1. *Researchers really don't have a robust estimate of how many "Joes," "Marys," and "Michaels" there are.* Yet simply knowing how many of these people there are would have serious public policy implications. The types of interventions necessary to help Joe, Mary, and Michael vary widely and include the option of doing nothing at all. But the implications of the interventions (or doing nothing) cannot be calculated unless we know how many people we're talking about.
2. *Depending on when researchers selected samples for study, Joe, Mary, and Michael may be "included" in studies of social stratification and labor force activity or "excluded" instead.* The implications of including or excluding them as individuals are trivial. The implications of including or excluding all others like them may not be.[1]

Consider three simple examples using each of our workers. If, as an unemployed steelworker, Joe is included in a study of earnings inequality in the 1980s, there is a good chance the estimate of inequality will be larger with Joe in than with Joe out. If Mary were included in a study of access to full-time employment, how would she be classified? She is in continuous contact with the temporary help agency. Some of her lack of employment is voluntary (those times when she decides not to take an assignment). At other times, Mary would like to work more hours but cannot. Michael represents an especially serious conundrum for researchers. The Michaels of this world are hard to find. On top of that, it is extremely difficult to classify their labor market activity. What does a researcher do with people whose job prospects are so poor that they can quit jobs with impunity (or get fired) without penalty?

Policy analysts gear their policies toward each of our workers depending (in part) on who they believe is most deserving and how many deserving citizens they think are out there. Most discussions of welfare reform and "deadbeat dads" who don't pay child support are geared toward men like Michael. Most discussions of what to do about "structural unemployment" are geared toward workers like Joe. The growing plight of temporary workers without established sets of job rights describes a situation like Mary's. If opportunities for all three are declining, this might explain changes in the popularity of other policies as well (for example, the growing hostility toward welfare and AFDC). These simplified scenarios point to the growing importance of studying aggregate labor market trends.

WHY STUDY LABOR FORCE ACTIVITY AT ALL?

The study of labor force activity can shed light on many current public policy issues in addition to answering basic fundamental questions that interest stratification researchers.

The biggest reason for studying labor force activity, and especially the study of labor force behavior over time, is that *full-time, full-year employment of most capable adults can no longer be taken for granted.* Clogg has argued in earlier work (1979) that full-time, full-year work never occupied more than 60% of the post-World War II labor force. This is an astonishing statistic when compared to the overriding social norm that able-bodied adults will work and that steady, full-time year-round employment will be used as a standard marker of success in adult life. The mere recognition of these statistics suggests that stratification researchers should be interested in studying who gets full-time/full-year work, what influences some people to attain full-time/full-year work and why, the correlates of full-time/full-year work, and the permeability of the boundary between full-time/full-year work and other levels of labor force attachment.

A second (and closely related) reason for studying labor force activity is because *most of the current policy debates about social inequality and what to do about it revolve around access to full-time/full-year employment.* The path toward a solution to this policy debate involves (1) methodologies for distinguishing full-time/full-year work from underemployment and unemployment, (2) assessing whether there are systematic biases in the allocation of full-time/full-year work, and (3) assessing whether full-time/full-year work is sufficient to substantially reduce poverty for those who have access to it. None of the current treatments of this policy debate take the relationship between any of these dimensions of labor force activity for granted. Instead, dimensions of labor force activity have to be measured and their conceptualizations defended against alternatives.

A third reason for studying labor force activity is that *it has the potential to shed considerable light on the issue of working poverty.* There is a growing debate in most developed nations (and many less-developed ones) about the existence and prevalence of working poverty. Working poverty reflects a situation where those employed in the labor market cannot secure high enough wages to avoid being classified as poor. Working poverty is an especially egregious situation in societies committed to equal opportunity because it violates deeply held equity norms. In most nations and cultures, people expect to receive rewards in rough proportion to their efforts in securing desired goals. Working poverty does serious violence to these norms and, in places where it is pervasive, can undermine normative commitments to work. Virtually no policy analyst in the United States believes that working poverty is desirable. Instead, debates focus on the existence and prevalence of working poverty. Those who study labor force activity are at the heart of this issue.

A fourth reason for studying labor force activity is that *such activity provides a ready way to gauge the integration of minorities and immigrants into labor markets.* Few would argue that minorities and immigrants are incorporated into dominant cultures if their labor force activity differs drastically from that of most people in the dominant culture. At a minimum, if one wants to argue that such groups are fully integrated, systematic differences in labor force activity have to be explained. Further, without an understanding of the full spectrum of labor force activity (from complete inactivity to full-time/full-year employment), one will have difficulty understanding why the stratification system looks the way it does in spite of the best efforts of politicians and social activists to change it.

A fifth reason for studying labor force activity is the growing recognition that *labor market activity has "voluntary" and "involuntary" characteristics that are melded together in complex ways* (as Mary's story above makes clear). In spite of the massive changes in the post-World War II economy of most developed nations, there is a growing recognition that people are choosing to be in contact with the labor market in an intermittent fashion and that such choices are not simply forced on them by necessity. Instead, most labor market decisions involve some voluntary components ("I could work full-time, but I don't want to because . . . ") and involuntary components ("I would like to make more money, but I can't seek training to increase my earnings because I can't afford it."). This follows from the statistics that Clogg provides. It would be hard to believe that the 40% of those who are not in the labor market full-time/full-year *all* want full-time/full-year jobs at any one time. Clearly some do and some do not. Understanding who belongs to which group, and why, are important components of understanding the structure of opportunity in labor markets.

A final reason for studying labor force activity is that *it varies across areal units and over time.* These differences have very real implications for public policy across geographic areas, and may help to explain the economic prosperity and hardship of these areas. We can think of two examples that illustrate geographic differences and their implications. The American states are increasingly in competition with each other for new capital investment and competitive advantage in luring and keeping business activity within their borders. Some states attempt to compete by offering cheap factor inputs such as low-cost labor. These states are likely to have workforces with low human capital or (perhaps more desirably) have unused skills, so that it is possible for incoming businesses to capture productivity gains from hiring underutilized workers. Other states attempt to offer high-productivity workforces with considerable human capital and considerable commitment to work. These states claim to offer scarce skilled talent that is difficult to find elsewhere and that can offer competitive advantages to companies needing highly skilled factor inputs. Both types of states will differ widely in what their labor force activity looks like.

As a second example, national employment policies will produce interesting comparisons of labor force activity. Nations committed to extensive regimes of workplace rights often tolerate high levels of unemployment and labor force inactivity for the sake of creating smaller numbers of high-quality jobs. Other nations committed to "full employment" policies often create thousands of jobs of varying quality, skill, and duration. These patterns of labor force activity have very different consequences for how their stratification systems work.

These are just a few of the many reasons why studying labor force activity is especially salient right now. Our book seeks to provide an overview of current issues in the study of labor markets by using labor force activity as a guiding concept.

A BASIC INTRODUCTION TO THE ISSUES EXAMINED IN THIS BOOK

In this introduction, we briefly review the contributions of Clogg and his associates to the study of labor force composition, underemployment, occupational mismatch, social mobility and socioeconomic attainments. Section II of the book discusses the uses and transformations of the Labor Utilization Framework. Section III focuses on the analysis of trends in labor force activity. In Section IV we review research on social mobility and socioeconomic attainments. In Section V we revisit some of the earlier work, and bring the analyses up-to-date through the mid-1990s. Finally, in Section VI we discuss a future research agenda born out of the culmination of the research discussed here.

The Labor Utilization Framework

The early 1970s was a time of great intellectual ferment in the study of social stratification in the United States. Clogg's work on underemployment was shaped by the 1974 recession, then the worst since the 1930s. Numerous scholars were pointing with increased dissatisfaction to models of the labor market that did not take into account the basic structure of labor market opportunities. Conventional status attainment treatments of occupational acquisition appeared far too automatic for a turbulent economic environment where there seemed to be an increasingly dubious connection between educational attainment and job opportunities for recent high school and college graduates. Further, the incorporation of women and minorities into the labor force appeared to be stalled. Several prominent studies were pointing to the increased work alienation and skill decline of many jobs in the labor market and warned of further problems ahead if current trends in the direction of increased automation and managerial control continued (see Kerr & Rostow, 1979).

It is in this context of anxiety concerning future labor market opportunity that Clogg expanded on the Labor Utilization Framework (LUF) initially proposed by Hauser (1974, 1977) and Sullivan (1978). The LUF was a general-purpose system for describing labor force activity. It shifted discussion of labor market opportunities away from unemployment toward a series of underemployment categories that reflected less attachment to the labor force than full-time, year-round employment. These categories implied that the labor force contained groups of people whose skills and talents were not adequately utilized in addition to the unemployed and those out of the labor force. The categories of the LUF served to differentiate the reasons for the underutilization of individual skills.

The LUF, as defined in *Measuring Unemployment: Demographic Indicators for the United States* (1979, pp. 6–10), included the following categories:

- *Those not in the labor force,* who are economically inactive and not seeking employment in the labor market
- *The subunemployed,* those who would be classified as "discouraged" workers. These workers are only marginally active in the labor market and do not look for jobs because of their lack of prior success
- *The unemployed,* as defined by the U.S. government
- *The part-time employed,* employment that reflects inadequate hours contributed to the labor market
- *Low income underemployment,* employment that does not provide a wage greater than or equal to 125% of the poverty line (as defined by the U.S. government)
- *Educational mismatch,* possession of educational qualifications (years of education) that are more than one standard deviation above the occupational average
- *Adequate employment,* those who do not fit in the above categories

As a measurement framework, the LUF was a response to several inadequacies in traditional labor force analysis. Chief among these was the belief that a substantial amount of inequality in labor market opportunity (and subsequent social stratification) was being hidden by the traditional labor force categories of "employed" and "unemployed." Since full-time, full-year employment could no longer be taken for granted, access to full-time, full-year employment was (potentially) a major factor in the production of social stratification. Differences in socioeconomic standing produced by adequate and inadequate employment could be as great or greater than those produced by differences in skill, product market position, authority or autonomy among the adequately employed. This initial work with the LUF developed into a research agenda devoted to identifying socioeconomic strata based on contact with the labor market.

Characterizing the Class Organization of Labor Market Opportunity

We close Section II by briefly touching on some earlier treatments of the
LUF that are outlined in a paper in *Sociological Methods and Research* (Clogg
1980) titled "Characterizing the Class Organization of Labor Market Opportunity." The purpose of this work was to show how latent structure methods
(then relatively new to the social sciences) could be applied to a substantive
sociological problem, providing insights that were not readily available through
other avenues.

Clogg uses four of the seven categories from the original LUF—unemployment, part-time employment, low income underemployment, and educational mismatch—as dichotomous observed variables in the latent class analysis. These four variables reflect the *time spent* in employment, the *earnings*
derived from work, and the *utilization* of worker skills.

The initial analysis was completed using the 1969–1973 Current Population Surveys, comparing loglinear results with results using latent class models. The results of the loglinear analysis yielded a very important substantive
conclusion. Workers who were currently unemployed or employed part time
usually had substandard earnings the prior year. Further, much of the marginality experienced in the underemployed portion of the workforce was related to lack of earnings and educational mismatch.

Results of the latent class analysis revealed that the labor force may be
accurately characterized by two latent classes: a marginal class with a relatively high risk of underemployment and a nonmarginal class with a relatively
low risk of underemployment. Here, the sensitivity and usefulness of the LUF
measure and of the latent class analysis are demonstrated in our ability to
compare the contribution of each type of underemployment to the marginal
labor force class over time.

Comparisons from 1969 to 1973 revealed a number of trends that would
probably not have come to light using other measures or methods:

- *Approximately 25% of the labor force falls into the marginal latent class and
  75% falls into the nonmarginal latent class.* These percentages are relatively stable from 1969 to 1973.
- While there is relative stability in the overall size of the marginal and
  non-marginal labor force classes, *there are noticeable fluctuations in the
  relative contributions of some types of underemployment to the marginal
  and non-marginal labor force classes.* Low-earnings underemployment
  remains stable at 38 to 40% of the marginal class. The biggest change is
  in the unemployment category, which fluctuates from 14.1% of the
  marginal class in 1969 to 22.2% in 1971.
- *Almost all of the underemployment for this time period can be found in the
  marginal labor force class.* Educational mismatch, the exception to this, is

greater in the nonmarginal labor force class, though educational mismatch grows steadily in both classes.

The substantive significance of this initial work comes from the finding that the labor force can be classified into two groups. The marginal group experiences intermittent employment at relatively low earnings. The nonmarginal class is only at risk of mismatch and experiences other forms of underemployment rarely, if at all. These categories, then, could be classified as separating the labor force into those who experience structural underemployment (those in the marginal class whose experiences of underemployment are a permanent feature of their working life) and those who experience frictional underemployment (those in the nonmarginal class whose experience of underemployment is temporary and tied to their higher expectations of labor market outcomes).

Trends in Labor Force Activity

In Chapter 3, we turn our attention to the analysis of trends in labor force activity. The analysis of trends in labor force participation and underemployment is motivated in part by the observation that women and minorities suffer from greater underemployment than do white men, and that the discrepancies in labor force participation between groups have grown during the 1970s and 1980s. Clogg's (1982a) initial analysis of trends in labor force participation compares labor force participators with non-participators. This work led to some interesting, and nonobvious, conclusions about trends in labor force participation among women and racial minorities.

First, in the 1969 to 1979 period, labor force participation is found to be higher for younger cohorts (compared with older cohorts) of nonblack males and females and for black females. For black males we see the reverse as older cohorts are more likely to participate in the labor force than younger cohorts. Although these cohort patterns are expected for nonblack females and black males and females, the patterns for nonblack males were unexpected. We address this in more detail in Chapter 3.

Second, the independent effects of age, period, and cohort are necessary to produce an adequate explanation of labor force participation trends. Moreover, cohort and period effects were shown to have countervailing influences on labor force participation.[2] For example, for black males, while the cohort effects act to reduce participation with each new cohort, the period effect acts to increase participation with each new year. For the other three race–sex groups, these effects are reversed.

Clogg and Sullivan (1983) develop further the analysis of time trends in labor force participation by examining trends in the LUF measure from 1969 to 1980. Their analysis sets out to examine the practical utility of the LUF against more conventional measures of unemployment. Because unemployment is a component of the LUF, comparisons across underemployment categories can be made easily.

Their results present a fairly clear description of where there has been change in labor force utilization over the decade of the 1970s. Using conventional measures of unemployment, one would have concluded that there has been (at worst) a cyclical change in labor force participation with unemployment varying between 3 and 9% of the labor force, reaching the latter figure in only one year.

However, an examination of the remaining LUF categories leads to much different conclusions about labor force trends. For example, there is a marked increase in educational mismatch during the 1970s. While mismatch increased among both males and females, the relative gender difference in mismatch remained constant over the decade. Nevertheless, males appear much more susceptible to mismatch than females and, by 1980, fully 17% of males in the labor force were mismatched.

Their results comparing blacks and nonblacks are also interesting and lend some support to prominent sociological literature on differences in the economic inclusion of blacks, especially the work of Wilson (1987). Specifically, blacks are much more likely to be unemployed or subunemployed than are nonblacks. The black to nonblack unemployment ratio remained constant (at about 2) and the subunemployment ratios increased over this time period from 2.6 to 3.9. Moreover, unemployment for blacks rises rather dramatically over the decade, from 6.2% in 1969 to 12.3% in 1980.

However, income-related underemployment for blacks dropped from 15% to 9% and the percentages for nonblacks remained relatively stable. And while mismatch rose substantially for both blacks (5.7 to 11.8%) and nonblacks (8.1 to 14.4%), the black to nonblack ratio remained remarkably constant given these increases. Overall, adequate employment for both blacks and nonblacks dropped almost 10% over the decade, while the black to nonblack ratio remained fairly constant.

We conclude Chapter 3 with the analysis of changing demographic composition and its effects on underemployment (Clogg & Shockey, 1985). More specifically, this work visits the question of whether underemployment rates should be adjusted for the demographic composition of the work force when making long-term comparisons in the rise and decline of labor market opportunity.

Without question the changes in the demographic composition of the labor force since the 1950s has been (and differences in unemployment across age groups are) considerable. The age composition of the labor force was altered in the 1960s by the entrance of baby boom cohorts, and the unemployment rate among 16- to 24-year-olds is two to four times higher than the unemployment rate for 35- to 40-year-olds. There were also substantial changes in the labor force activity of females and normative changes in expected labor force activity as well. However, the only change in the 1970–1980 decade that clearly favored an increase in unemployment was the greater representation of workers aged 20–34. Otherwise, declines among workers over 55 represented declines in age groups with higher unemployment rates compared with all but the youngest labor force entrants, so the net effect of changes in the age composition of the labor force on unemployment during the decade of the 1970s would appear to be close to zero.

Clogg and Shockey show the limited effects of compositional shifts on underemployment trends by comparing crude and adjusted rates across LUF categories. These results show that changes in demographic composition (and adjustments made to reflect them) make little or no difference in the conclusions one would draw about changing underemployment trends. In no case is the difference between the crude and adjusted rates substantial. This result suggests that the comparison of aggregate underemployment trends is acceptable and that underemployment rates across categories do not need to be adjusted for age and sex composition, at least for this time period.[3]

Social Mobility and Socioeconomic Attainments

In Chapter 4, we focus on the line of research oriented toward analyzing social mobility and socioeconomic attainments. We begin that section with the work of Clogg and Shockey (1985), a piece that elaborates in fairly straightforward terms a baseline approach to modeling mobility tables, followed by a description and critical comparison of Hauser's layers model and Hope's modeling approach.

First, the authors describe in detail the loglinear models for independence (I) and quasi-independence (QI), and how these relate to mobility table analysis. From this they launch into a basic description of Hauser's levels model, and how this builds on the I and QI models. With reference to three primary criteria for judging the models—(1) consistency with sociological theory, (2) development in accord with accepted statistical principles, and (3) usefulness in comparative research—both the strengths and the weaknesses of Hauser's levels model and Hope's modeling approach are discussed.

We next turn to the work of Clogg (1981) that further highlights the utility of the latent class model in sociological research. Taking three classic mobility tables, he discusses how a complex statistical method such as latent class analysis may be used to shed important new light on social mobility processes. Two latent classes are viewed as probabilistic focal points to which origins flow and from which destinations are attained. Transition rates are then decomposed into probabilities related to the flow into and out of these latent classes.

To extend the usual latent class approach to handle the high density of cases on the diagonal (which takes into account the fact that large numbers of people do not change occupations), Clogg then develops what he calls quasi-latent structure models. These quasi-latent structure models are variants of mover–stayer models where the diagonal cells are treated as deterministic stayer classes. For the three mobility tables examined, the quasi-latent structure models fit these data well. Clogg constructs indices of immobility from these quasi-latent structure models, indices superior to prior attempts because they are derived from models that fit the data.

We then move to work focusing on socioeconomic attainments, specifically the work of Clogg, Eliason, and Wahl (1990), Clogg and Eliason (1990), and Clogg, Ogena, and Shin (1991). This work demonstrates that individual labor force behaviors significantly affect individual socioeconomic attainments. Prior work in this area viewed attainments of socioeconomic status through the lens provided by the work of Peter Blau, Otis Duncan, William Sewell, and Robert Hauser. What Clogg and his associates showed was that a more complete picture of the attainment process would be achieved by including individual labor force behavior. Specifically, Clogg and associates discuss an expanded LUF measure cross-classified with labor market experiences over the course of a year. From this, scales were developed measuring different dimensions of labor force behavior. These scales were then added to classical models of socioeconomic attainments and were shown to significantly affect those attainments.

Taking the original specifications found in Featherman and Hauser's (1978) classic work using the OCG-II sample, Clogg et al. (1991) show that both labor force position and labor market experience scores have significant effects on status attainment, and they provide a considerable increase in the variance explained in models of log-earnings. The inclusion of position and experience scores in the Featherman/Hauser models reduces the effect of education on log-wages. Other methods of correcting for labor force marginality and sample selectivity do not capture the same information about labor force behavior and experience as measures of past labor market experience and current labor force position.

Trends through the Mid-1990s

In Chapter 5, 7 and 8, we turn our attention to updating the analysis of labor market activity through the 1990s. This analysis is constructed with an eye toward answering several questions: (1) Has there been a permanent shift in the composition of unemployment and underemployment categories as a result of deindustrialization and corporate downsizing?, (2) Are standard LUF categories (combined with measures of prior labor market activities) able to capture what many believe to be an increasingly dynamic and unstable labor market? (3) What deficiencies does this analysis reveal that could be corrected using other conceptual tools available in the sociology of labor markets? The analysis is conducted using March CPS data from 1982 to 1995, and sheds some light on the durability of the conceptual tools developed in this book.

A Future Research Agenda for the Study of Labor Markets

In the final section of the book we outline the future implications and research agenda for students of labor markets using the tools we have described in prior chapters. We believe that this prior research has provided a solid foundation for new and exciting research on a variety of topics that are of interest to sociologists and policy analysts. In particular, we think that future research based on the conceptual apparatus, measures, and models discussed in this book can shed considerable light on issues such as labor force attainments, labor market segmentation, and the effects of different employment and income policies on labor utilization. In a cross-national context all of these issues have rich policy implications in rapidly diversifying developed and developing nations.

We hope at the end of this book that you are as excited about the possibilities (and as understanding of the limitations) of our conceptual framework and the study of labor force activity as we are. If this volume does nothing more than promote further study of labor force behavior and the rapidly changing labor markets in the developed world, then our mission will have been realized.

NOTES

1. Whether excluding different classes of workers with different employment histories will actually affect one's research conclusions depends in part on the sample size. Excluding classes of labor force participants (or nonparticipants) may induce sample selection bias into an analysis, and one way of knowing this is to compare results with results derived from more inclusive samples.

2. Further, models containing an age-by-period interaction appear most theoretically satisfying in lieu of generally adequate model fits.

3. The only measure for which adjustment makes a modest difference is the mismatch measure. Adjusting for the composition-by-year and composition-by-year-by-LUF interactions yields estimates of mismatch that are higher in 1969. Compositional changes adjusted using a stable population model produce more dramatic effects, especially for measures of unemployment, but the analysis does not discuss extensively the favorability of using stable population comparisons in calculating rates of unemployment change over time.

2

An Introduction to the Labor
Utilization Framework

INTRODUCTION

Since the 1930s most Western industrialized nations have used unemployment statistics to assess the general condition of the labor market. Policymakers
approached the construction of unemployment statistics by separating the
working age population into two groups. The first group included those who
voluntarily did not work for pay or profit. These people were referred to as the
economic inactive population. The second group included those who did work
for pay or profit as well as those who were out of work but actively seeking a
job. Researchers assumed that the loss of employment and the inability to
find employment represented a failure in the mechanisms that allocate employment opportunity. People in this second group were exposed to the "risk" of
employment and unemployment so this segment of the population was referred
to as the *economically active population* or the *labor force*. Virtually all labor
force statistics since the 1930s have been based on the separation of the economically active from the economically inactive population (see Hauser 1949).

Researchers used this approach to construct the *unemployment rate*. The
unemployment rate was the number of economically active people without
employment expressed as a percentage of the total labor force. Changes in
the unemployment rate reflected the changing economic conditions of labor
markets. Researchers could describe the relative position of women and
underrepresented groups in the labor force through cross-sectional compari-

*This chapter is taken and modified from *Measuring Underemployment*, Academic Press, 1979.

sons of unemployment rates. The unemployment rate has been one of the most influential statistics for social science and public policy and has made and unmade political fortunes.

The underlying premise of this book is that the unemployment statistic is no longer an adequate indicator of the economic and social condition of the labor market. Instead, we argue that information on underemployment is needed to describe the quality and quantity of work provided to the economically active population. Thinking about underemployment requires us to consider new concepts for describing the structure of the labor market, and new methodological tools for summarizing the complexity of labor markets that vary radically in the quality and quantity of work they provide.

The conceptual groundwork for our study is contained in a resolution adopted by the Eleventh International Conference of Labor Statisticians in 1966. The resolution (see International Labour Office, 1976, pp. 33–34) reads:

> Underemployment exists when a person's employment is inadequate in relation to specified norms or alternative employment, account being taken of his occupational skill (training and experience). Two principal forms of underemployment may be distinguished; visible and invisible.
>
> (1)Visible underemployment is primarily a statistical concept directly measurable by labour force and other surveys, reflecting an insufficiency in the volume of employment. It occurs when a person is in employment of less than normal duration and is seeking, or would accept, additional work.
>
> (2) Invisible underemployment is primarily an analytical concept reflecting a misallocation of labour resources or a fundamental imbalance as between labour and the other factors of production.
>
> Characteristic symptoms might be low income, underutilization of skill, low productivity. Analytical studies of invisible underemployment should be directed to the examination and analysis of a wide variety of data, including income and skill levels (disguised underemployment) and productivity measures (potential underemployment).

The conference participants pointed to the need to measure both the quality of work available to the economically active population in addition to the sheer quantity of work available. The dimensions of *time spent* in employment, *income* derived from work, the *productivity* of work, and the *skill utilization* of workers were necessary components in the measurement of work force "underutilization" or underemployment.

The time dimension of underemployment has received the most attention. The almost worldwide acceptance of unemployment rates as the standard measure of the health of labor markets is testimony to that. Most labor force surveys are conducted with the primary objective of measuring the time dimension of underemployment. The income, productivity, and skill utilization dimensions are "invisible" characteristics of the quality of work—they are not routinely collected and distributed as signs of labor market health, they are more difficult to measure, and researchers and analysts knew very little about their typical patterns and prevalence until very recently.

To shift the focus of researchers and policy analysts toward a broader set of employment indicators, we offer the Labor Utilization Framework (LUF) as a means of measuring and tracking underemployment in the labor market.

THE LABOR UTILIZATION FRAMEWORK

The LUF was first proposed by Hauser (1974, 1977) to deal with deficiencies of the unemployment measure when examining the labor markets of developing nations. Studies by Myrdal (1968), Turnham (1971), and others summarized the failure of Western definitions of unemployment for developing societies and concluded that the volume of unemployment actually measured was only a small fraction of the "subproductive" or "underutilized" labor force.

The LUF was designed to remedy the deficiencies in the traditional labor force approach. At the time of this writing, the LUF has been successfully applied in several developing nations, including Thailand, Indonesia, Singapore, Malaysia, the Philippines, Taiwan, Hong Kong and South Korea. The natural question to ask is whether the LUF, or some close relative to it, can be used with equal justification in developed societies like the United States.

Researchers recognized the deficiencies of unemployment statistics soon after they were adopted. Robinson (1936) described the existence of a substantial amount of "disguised" unemployment that could not be uncovered by the now conventional approach. Some years later, after nearly two decades of experimentation with the Current Population Survey, the National Bureau of Economic Research (1957) sponsored a conference on the meaning and measurement of unemployment. It was evident to the participants of this conference that the meaning of unemployment statistics was still unclear even after extensive experimentation and comparative analysis. Several of the participants in this conference (e.g., Gertrude Bancroft and Albert Rees) suggested slightly different measurement schemes, although general proposals for the measurement of underemployment were conspicuously lacking. In the early 1960s, President John F. Kennedy commissioned a special committee to appraise unemployment statistics, the first indication of governmental dissatisfaction with labor force measurement procedures. The Gordon Committee did much to foster a public awareness of the deficiencies in the labor force approach. Recommendations for moderate revisions in definitions were made, but few incentives were given in this policy directive for radical change (see U.S. President's Committee to Appraise Employment and Unemployment

Statistics, 1962). Proposals stemming from the Gordon Committee were largely aimed at obtaining sharper measures of unemployment itself, sharper measures of part-time unemployment, and minor redefinitions of the economically active population.

The controversy over unemployment measures resurfaced most vividly in the events that led to the passage of the Comprehensive Employment and Training Act of 1973 (CETA). The CETA legislation contained a mandate for governmental statistical agencies to develop measures of underemployment and constituted a dramatic change of direction in the general activity of labor force measurement. In 1977, President Jimmy Carter commissioned another committee of experts headed by Sar Levitan, partly with the purpose to appraise the various proposals for the measurement of underemployment. The LUF was not one of the frameworks being evaluated by this important committee (Sullivan & Hauser, 1979).

The controversy about conventional measures of unemployment essentially hinged on two issues. The first was the need to further refine the measures of unemployment, perhaps including a refinement of the labor force definition itself. Certainly this has been the dominant theme in the controversy about the unemployment statistic from the Depression through the 1960s. The second issue is the need to develop supplemental labor force indicators, including measures of underemployment or underutilization. We focus on this second issue by attempting to develop an operationalization of underemployment that will be useful to researchers and policy analysts.

The five forms of underemployment we use apply to those who are members of the economically active work force. Persons are members of the active work force *if they are working for pay or profit or actively seeking work*.[1] One longstanding criticism of unemployment statistics is that they ignore "discouraged workers." Discouraged workers are marginally active workers who are out of work but who refuse to actively seek work because of their unsuccessful past efforts to find a job. Discouraged workers fluctuate in number with economic growth and recession. Some charge that they should be included in the measurement of the underemployed, since their numbers depend on many of the same economic fluctuations that drive unemployment (see Flaim, 1972). Nevertheless, the standard approach has been to exclude them from the universe of the economically active, which renders them invisible to unemployment statistics.

We cannot isolate discouraged workers exactly in accordance with official definitions, but we do use a method that approximates the notion of discouraged workers. We refer to people identified by our method as the *subunemployed* to reflect the apparently greater severity of underemployment for these people relative to the unemployed.

Our conceptualization uses standard, official definitions of *unemployment*.

We do this to maintain some consistency with past practices. We also can compare the relative size of our other underemployment categories with official definitions of unemployment to make conclusions about what conditions official definitions of unemployment ignore.

We also rely on working time to estimate *involuntary part-time employment*. This form of underemployment refers to workers who are working less than a full-time workweek because they cannot find a full-time job. All three measures of underemployment have been routine components of labor force reporting for several decades.

The obvious common denominator of unemployment, subemployment, and involuntary part-time employment is the concept of *work time lost*. If the working hours provided to the labor market are all of the same quality and workers are adequately using the skills and human capital they possess, then work time lost is a critical dimension of underemployment.

Bowen and Finegan (1969) illustrate the preoccupation with the time lost dimension of underemployment. In their study, the fullness of participation in the work force is indicated only by the hours worked in a week or by the weeks worked in a year; the implicit conception of adequate employment (or of underemployment) is centrally linked to the amount of time worked. Most social scientific work concerning underemployment has focused almost entirely on the time dimension of underemployment, to the exclusion of the other dimensions or forms.

The obvious criticism of approaches with an exclusive focus on the time dimension of underemployment is that time-divisible work units in most economies are not exactly the same. Clearly, time spent in employment can be further classified along dimensions of income derived from work, the productivity of work, and the skill utilization of workers. By looking at these dimensions of employment, we can gain knowledge of how the labor force is productively used and how the tangible rewards of the labor market are distributed.

In addition to indicating the quality of time spent in employment, our invisible underemployment measures should also measure the changes that occur in the balance of supply and demand in the labor market. Time lost by the labor force may not reflect changes in supply–demand schedules but may be inordinately influenced by, say, government controls and trade union pressure. The economic exigencies that presumably would have forced employers in a free labor market to decrease the hours of employment extended to workers may now instead force a lower effective wage rate, a stifling of worker productivity, or a shifting of skilled labor to unskilled positions, all while maintaining the same number of time units of work.

Some workers in this new labor force may be working full time and hence be counted as "adequately employed" workers by the standard methods, but their wages may not be sufficient to provide them with adequate income. These workers are referred to in our study as *low- income underemployed*. We consider a measure of low-income underemployment that is comparable to the standard labor force approach and does not abandon the more usually visible underemployment measures at all.[2]

Another kind of invisible underemployment occurs when workers are inadequately employed because their skills—their accumulated fund of human capital—are considerably greater than the skill requirements of their jobs. These people are referred to as *mismatched* in our study. Hauser's innovation in the "labor utilization framework" is the method of measuring invisible underemployment in general, and the mismatch form in particular.

These underemployment forms may occur singly or jointly. A worker who is mismatched may be temporarily unemployed and so on. Following Hauser's conventions, we can define these forms in a way that makes them mutually exclusive. People of working age may be classified into seven different statuses, including the statuses that denote the five underemployment forms of this study:

1. *Not in the labor force* (or economic inactivity)
2. *Subunemployment* (our proxy for "discouraged worker" status)
3. *Unemployment*
4. *Part-time employment*
5. *Low income underemployment*
6. *Mismatch*
7. *Adequate employment*

The conventional labor force is composed of people classified in any of statuses 3 through 7, and statuses 1 and 2 taken together denote the usual not-in-the-labor-force category of official reports. Since our subunemployment category and proxy for discouraged workers contains people thought to be marginally economically active, we operationally define a modified labor force as including people in statuses 2 through 7. These seven labor force statuses produce a classification that is much richer than conventional classifications. They also allow us to focus on inequality in the basic rewards available in the labor market, that is, access to full-time, year-round employment at an adequate wage that sufficiently uses workers' skills.

In summary, there are three main dimensions of underemployment—*work time lost, income deficiency*, and the *mismatch* of workers' skill attainment with required job skills—and they all relate directly to three logically related types of labor market rewards.

A CRITIQUE OF THE LABOR UTILIZATION FRAMEWORK

Researchers have criticized the LUF on several fronts. They have argued about the appropriate cutoff for the measurement of low-income underemployment. We take these thresholds from the Social Security Administration's recommendations for determining the poverty status of households. Some have charged that the usage of poverty thresholds implies that the measure of low-income underemployment is actually an indicator of poverty or need, not necessarily underemployment in the labor market. But our poverty cutoffs were adjusted so that they could be applied to assess the deficiency of work-related income on a per-worker basis, an approach that is quite different from one emphasizing the actual poverty of households. The LUF uses the poverty thresholds merely as benchmark criteria for measuring low-income underemployment, but other criteria could be used as well. The important idea in any attempt to measure such a concept, however, is the application of uniform standards. As different income cutoffs would lead to very different absolute levels of low-income underemployment, the absolute number of low-income underemployed workers is difficult to interpret by itself. We would prefer to look at relative comparisons over time or across groups.

Another possible problem involves the use of completed years of formal education as a proxy for the skill attainment of the worker. Our technique does not take into account the obvious fact that educational inputs vary in quality and contribute in different ways to the real skill development of different workers. The skill level of two workers with the same completed years of schooling can differ solely because of differences in the quality of their education. Our use of educational attainment as a proxy for skill level does not take into account the obvious influence of on-the-job training or skill accumulation through experience, both of which are important vehicles for raising the stock of human capital. Nor does our measure take into account "credential inflation" or other mechanisms that would produce a secular increase in the apparent skill-level of jobs apart from whether those skills were actually used on the job.

These criticisms pertain to the way in which the forms of invisible underemployment are measured with existing data. They do not attack the need to measure the prevalence of the several forms but rather the specific techniques used. Other criticisms are more harmful and will be much more difficult to remedy in future attempts to measure underemployment. Most of these criticisms pertain to the ambiguity of defining and measuring the various forms of underemployment for certain kinds of workers. We can think of at least three categories of workers for whom it is extremely difficult to define, let alone measure, the existence or nonexistence of underemployment.

1. *Unpaid family workers,* found in certain types of small businesses, clearly represent a problem in the definition of the low income form of underemployment, as no money income is received by these workers. Unpaid family workers constitute 1–2% of the U.S. labor force, and women and nonwhites are disproportionately represented in this group. Any other workers who receive remuneration for their labor in a manner that is difficult to assign a

monetary value to are also subject to the same ambiguous classification. Many such workers report having negligible income and would be counted as low-income underemployed in most labor force surveys. The low-income form of underemployment as used in this study is clearly the most appropriate to the case where a definite wage income can be measured.

2. *Agricultural workers* constitute a problem in the measurement of most of the underemployment forms. Much agricultural work is not easily defined in terms of time spent in employment, income received from work, or skill requirements, even in a time when agribusiness threatens to destroy the traditional household organization of agricultural work. The underemployment forms of our study are best suited to the measurement of labor force conditions in the manufacturing and service sectors. Agricultural workers comprise between 2 and 5% of the active work force in the United States.

3. *Secondary earners* generally pose a serious problem in the definition of underemployment. Primary earners typically have an attachment to the labor force that is of long duration, and their attempts to secure labor market rewards are more likely to be planned in reference to a long-term career. Secondary earners have a more transient attachment to the labor force, reflected in their multiple entries to and exits from the labor force during the life cycle. (Of course, the very transience of employment for secondary workers contributes to the advantage of the primary earners, a point repeatedly made by feminist analyses of labor markets, see England, 1992.) Secondary workers, by most definitions, belong to households, and these households are organized around a primary earner who is attempting to optimize his or her capacity to seek adequate employment. If several earners in a household were all optimizing their capacity to seek out adequate employment (in the sense of maximizing individual-level utility), then households (as we conceive of them) could have only a momentary existence. The necessary geographic mobility that accompanies a worker's optimal work-seeking strategy would dissolve households containing several workers in only a short span of time. But we know that some households with several workers apparently persist through time, in part because employment priorities have been negotiated satisfactorily. These priorities imply that different standards for ascertaining underemployment for each type of worker should be developed. Such standards would be necessary to reflect the apparently different severity of underemployment for these two types of workers. Technical definitions of secondary workers are changing as we speak, and there will almost certainly be an increase in the proportion of workers whose attachment to the labor force is transient, secondary, or dependent on the labor market experience of fellow household members. We expect the question of secondary workers to increasingly affect both men and women in the future as well.

The problem of measuring underemployment by any method thus far suggested in the literature is far from a trivial one when these kinds of workers are encountered. Currently almost one-half of the active work force in the United States are unpaid family workers (or workers very similar to unpaid family workers), agricultural workers, or secondary workers. Labor force indicators in general would be greatly improved if a method could be developed for dealing with these kinds of workers that did not abandon the conceptualization of different underemployment forms. This subject is currently overburdened with ideological overtones, especially with regard to the treatment of secondary workers, so no rational solution would seem to be immediately forthcoming. It is necessary that these considerations be kept in mind so that the statistical summaries of underemployment, especially of differential underemployment, will be treated cautiously in the remainder of this work.

A LATENT CLASS PERSPECTIVE FOR THE ANALYSIS OF THE WORK FORCE

In the logic of a latent class analysis of the labor force, we assume that there are a number of underlying labor force classes that govern the relationship among salient labor force attributes. That is, the placement of individuals in the conjunction of these salient labor force attributes is in large part governed by the (latent) labor force class in which they find themselves. However, we cannot observe these labor force classes directly. Instead, we infer class positions using information on the observed attributes. Thus, the latent class analysis provides a way to combine the information on these observed attributes into packages that represent distinct labor market segments. Members of different labor market segments, or classes, differ significantly in the opportunities afforded them by the labor market.[3]

We use four observed measures to make inferences about the class organization of labor market opportunity. Our measures are defined in terms of the labor market rewards of *time spent in employment, income derived from work,* and the *utilization of acquired worker skills.* Two of the measures are commonplace and do not require much commentary from us. Regarding time spent in employment, dichotomous measures of unemployment and involuntary part-time unemployment measure the inability to gain access to a privileged portion of the labor market (see U.S. Department of Commerce, 1967). Our third measure is a dichotomous income measure that is constructed by comparing work-related income from the past year with the "poverty threshold" income level appropriate for a worker of the same age, sex, residence, and household characteristics (see U.S. Department of Commerce, 1975).[4] If a worker's income is below this threshold, we classify the worker as "low-income underemployed." Our fourth measure assesses when a worker is "overskilled" for the job he or she currently holds. The completed number of years of education is used as a proxy for the skill level of a worker, and we compare this to the average number of years of education that other workers in similar occupations have. If a given worker has more education than the mean education level plus one standard deviation for his occupational group, then we categorize the worker as "mismatched," or overskilled for the current job. For our purposes we note that: (1) the underemployment measures are all dichotomous; (2) we can apply the criteria over time;[5] and (3) we can cross-classify the four measures since the measures of underemployment are not mutually exclusive (e.g., a part-time unemployed worker can be educationally mismatched regarding her or his current occupation).[6]

SIMPLE LATENT STRUCTURES APPLIED TO THE 1970 DATA

We present results for some latent class models in Table 2.1. In model H_0 we assume all individuals are in a single "latent class."[7] Model H_0 fails because the number of "latent" classes is greater than one. The implication here is that we should not assume that the labor force is composed of only a single class.

Given the special meaning of the low-income measure—defined with reference to the earnings experience over the past year—we ask next if the low-income measure itself can explain the association among the underemployment forms. In other words, can knowledge about workers' income in 1969 account for the association between the time and mismatch forms observed in 1970? Using H_0 as a baseline model providing an estimate of total associa-

Table 2.1
Chi-Square Values for Models Applied to 1970 Data

Model	Degrees of freedom	Likelihood-ratio chi-square L^2	Goodness-of-fit chi-square χ^2
H_0 (=M_1)	7	2233.05	3274.92
H_1 (=M_4)	4	36.75	33.11
H_2	3[a]	5.50	5.45
H_3	4	6.77	6.67
H_4	5	10.28	10.01
H_5	6	1853.44	2119.02
H_6 (=H_7)	3[a]	1.01	1.00

[a]Since one of the parameters (π_{12}^{EX}) was estimated as zero, we assign one more degree of freedom to these models. For justification for this convention, see Goodman (1974a).
Source: Clifford C. Clogg, 1980a, "Characterizing the Class Organization of Labor Market Opportunity." *Sociological Methods and Research*, Vol. 8, p. 253.

Table 2.2
Underemployment in the Low-Income and Adequate Income Parts
of the Labor Force

Underemployment form	Low-income Part (= marginal class)	Adequate-income part (= nonmarginal class)
Unemployment	17.4%	3.5%
Part-time employment	8.9%	2.0%
Mismatch	9.1%	10.8%
Percent of labor force in each class	9.4%	90.6%

Source: Clifford C. Clogg, 1980a, "Characterizing the Class Organization of Labor Market Opportunity." *Sociological Methods and Research*, Vol. 8, p. 256.

tion to be accounted for, we see that this hypothesis (H_1 in Table 2.1) accounts for fully 98% of the association between time and mismatch measures of underemployment. In short, the latent class structure of labor market opportunity is strongly defined by income classifications.[8]

To demonstrate how well the income indicator predicts the risk to other forms of underemployment, we present rates of unemployment, part-time unemployment, and mismatch for the "low-income" and "adequate-income" parts of the labor force in Table 2.2. We see that 17% of the labor force with low work-related income in 1969 was unemployed during the survey week in 1970. This unemployment rate is five times higher than the rate for workers with adequate income from work in 1969. Part-time unemployment is also extremely high in the low-income part of the labor force.

There is a remarkable persistence of marginality for those workers whose work-related income in the past year is substandard; clearly, the low-income part of the labor force is in a marginal labor force position by any definition of the term. When unemployment and part-time unemployment are considered together, 26% of those in the low-income portion of the labor force are at risk of either not working enough hours or of not finding work at all. Low-income work, then, seems to clearly reflect a central component of labor market segmentation in the United States.

There are several practical implications that follow from our results so far. First (as already noted) the transition between low-income employment and adequate-income employment seems to be a fundamental stratifying dimension of the labor market. Second, policy debates about the labor force attachment of low-income people appear to be on target, at least with regard to the general parameters of the debate. Do those with low-income employment simply work less than those with adequate-income employment or does the unstable nature of low-income work contribute to labor market marginality on other dimensions (inconsistent working hours, frequent layoffs, and so

Table 2.3
Underemployment Rates for the Marginal and Nonmarginal Labor Force
Classes (Two-Class Unrestricted Model)

Underemployment form	Marginal class	Nonmarginal class
Unemployment	23.6%	0.0%
Part-time employment	12.3%	.2%
Low income	34.0%	3.2%
Mismatch	8.0%	11.2%
Percent of labor force in each class	20.2%	79.8%

Source: Clifford C. Clogg, 1980a, "Characterizing the Class Organization of Labor Market Opportunity." *Sociological Methods and Research*, Vol. 8, p. 257.

on)? The analysis so far at least suggests that income and hours worked are related to each other (e.g., see also Mead, 1991; Wilson, 1996).

Our third model, model H_2, in Table 2.1 assumes that each of our measures of labor force marginality are fallible indicators of class. In this model the estimated risk of unemployment for the nonmarginal class is zero. Under model H_2 we see that unemployment is necessary, although not sufficient, to identify marginal workers, because all unemployed persons are regarded by H_2 as being in the marginal class. The fit of this model for the 1970 data is very good, with $L^2(H_2) = 5.50$; a latent structure consisting of only two classes is sufficient to describe the structure of opportunity for the labor force.[9]

The underemployment rates for each latent class from model H_2 are presented in Table 2.3. It is seen that this model determines one's class location on the basis of time spent in employment. The marginal class has a very high risk of unemployment and part-time unemployment, and the nonmarginal class has zero risk of unemployment and a very low (0.2%) risk of part-time employment. This model suggests that 20 percent of the labor force is marginally connected to the labor market and 80% is not.[10]

A SIMPLER LATENT CLASS STRUCTURE

In spite of the virtues of the last analysis we can make our two-factor description of the labor force simpler still. We now develop and estimate a model where we conceive of a single nonmarginal class whose labor force chances are constrained in such a way that only one kind of underemployment can occur to a nonmarginal worker at any given time (model H_6 in Table

2.1). These restrictions give rise to a six-class model, rather than a two class model. Nevertheless, this six-class model can be thought of as a reasonably simple kind of two-class model. Multiple underemployment is prohibited in a single latent nonmarginal class here, and that is all that distinguishes this model from the unrestricted two-class model H_2 presented earlier.

This model fits the observed data very well. The statistics in Table 2.4 allow us to characterize the two classes. For the nonmarginal class, the risks of unemployment and part-time unemployment are not zero but are certainly low enough to represent frictional or transitory interruptions in labor force experience. We see that the risk to unemployment is 35 times greater for the marginal class relative to the nonmarginal class, the risk to part-time employment is 17 times greater for the marginal class relative to the nonmarginal class, and the risk of educational mismatch is only slightly smaller in the marginal class relative to the nonmarginal class. Because educational mismatch only applies to people with some post-high school education, we see that some people with high educational attainment (and correspondingly high job skills) are considered marginal in this model. We note that 25% of the labor force is in the latent marginal labor force class under this last model.

We think our simplified model is most compelling because: (1) it prohibits multiple forms of underemployment in the nonmarginal class; (2) it produces positive but low risks of inadequate working time for the nonmarginal class (analogous to "frictional" unemployment), and (3) it places all people with low income into the marginal class. The income variable, because it averages an entire year's work experience, should play a central role in our characterization of labor market marginality.

Table 2.4

Underemployment Rates for the Latent Marginal and Nonmarginal Labor Force Classes (Taken from Model H_7, Table 2.1)

Underemployment form	Marginal class	Nonmarginal Class[a]
Unemployment	17.5	.5
Part-time employment	8.8	.5
Low income	37.5	.0
Mismatch	9.1	11.1
Percent of labor force in each class	25.2	74.8

[a]Multiple forms of underemployment are impossible in this model. The underemployment forms are quasi-independent for workers in the non-marginal class.
Source: Clifford C. Clogg, 1980a, "Characterizing the Class Organization of Labor Market Opportunity." *Sociological Methods and Research*, Vol. 8, p. 263.

TIME-PERIOD CHANGE IN THE LABOR FORCE VIEWED FROM THE LATENT CLASS PERSPECTIVE

In substantive terms, the structural parameters presented in Table 2.5 are also meaningful and serve to characterize the time trend in the labor force. In the data considered here, 1969 represents the most favorable labor market conditions and 1972 represents the least favorable labor market conditions of a recession economy. We see that the latent marginal class is 24.4% of the total labor force in the former year but that this proportion increased to 26.4% in the latter year, an increase of 2% in the proportion of workers in a marginal condition. For the marginal class the risks of unemployment and part-time unemployment fluctuate a great deal over time. In 1969 the marginal class had a 14% risk of unemployment, but in 1972 it had a risk of 22%—an increase of 8%. That is, the moderate overall increase in unemployment risk for the aggregate labor force 1969–1972 of (6.3% – 3.6% =) 2.7% was accompanied by a tremendous increase in risk of 8% for the marginal class. For the nonmarginal class, the corresponding increase in the risk of unemployment was a slight (0.8% – 0.2% =) 0.6%.

Table 2.5
Latent Labor Force Classes Across Time

Underemployment risk	1969	1970	1971	1972	1973
			Latent marginal class		
Unemployment	14.1%	17.5%	22.2%	21.6%	19.0%
Part-time employment	8.1	8.8	10.0	9.3	8.8
Low income	38.2	37.5	37.7	39.5	40.4
Mismatch	8.8	9.1	11.2	13.2	12.8
% in each class	24.4%	25.2%	25.6%	26.4%	25.2%
			Latent nonmarginal class		
Unemployment	.2%	.5%	1.0%	.8%	.8%
Part-time employment	.5	.5	.6	.6	.4
Low income	.0	.0	.0	.0	.0
Mismatch	10.7	11.1	12.3	12.7	13.8
% in each class	75.6%	74.8%	74.4%	73.6%	74.8%
		Likelihood-ratio and goodness-of-fit chi-square statistics			
L^2	2.55	1.01	3.20	13.52	5.88
χ^2	2.57	1.00	3.23	13.16	5.81

Source: Clifford C. Clogg, 1980a, "Characterizing the Class Organization of Labor Market Opportunity." *Sociological Methods and Research*, Vol. 8, p. 266.

The mismatch measure does not provide much information for distinguishing the two latent classes because the risks of mismatch are very similar in magnitude for the two classes over time. But it is the close similarity of mismatch risks between the classes that presents an enigma, because the risk of mismatch is so high in the marginal class. In fact, in the 1972 recession year mismatch was more likely in the marginal class than in the nonmarginal class. Because mismatch only applies to people with some post-high school education, and because mismatched people are the most skilled (i.e., the most educated) members of their occupations, it is puzzling (at first sight) that the risk to mismatch would be so high in the marginal class. We would expect that the most skilled or most educated in various occupational groupings would have negligible risk of unemployment, part-time unemployment, or low work-related income, and not be found at all in a marginal class. However, the observed characteristics of the labor force contradict our expectations. Our cross-classification of underemployment types shows that from 10.9% of the mismatched (in 1969) to 15.8% of the mismatched (in 1972) also experience some other form of underemployment (see Clogg, 1980, Table 1, p. 248). In view of the superior skill that the mismatched worker enjoys over counterparts in the same or similar occupations, such a high risk to the other forms for mismatched workers requires further explanation. These facts cannot be explained by a purely economic theory of competition in an unconstrained labor market, as such a theory would predict a definite advantage of the most skilled over the less skilled.

DECOMPOSING OBSERVED UNDEREMPLOYMENT INTO COMPONENT PARTS DUE TO THE DIFFERENT LABOR FORCE CLASSES

Table 2.6 presents calculations that decompose aggregate underemployment rates into components due to the marginal class and components due to the nonmarginal class.

From Table 2.6 we conclude that much of the observed time trend and most of the aggregate unemployment are due to unemployment among marginal workers. The aggregate rate of mismatch also can be decomposed into the two component parts in a way analogous to the above.

Although mismatch seems to be more important in the marginal class with the passage of time (with the exception of 1973), most of the aggregate level of and the time trend in mismatch are due to mismatch characteristics of the nonmarginal class. We could decompose the aggregate rates of the low-income and the low-hours forms by the above method, but because all low-income people are located in the marginal class, the trend in the rate of

Table 2.6
Observed Unemployment and Mismatch in the Marginal and Nonmarginal
Labor Force Classes

	1969	1970	1971	1972	1973
Observed unemployment	3.6%	4.8%	6.4%	6.3%	5.4%
Part due to the nonmarginal class	.2	.4	.7	.6	.6
Part due to the marginal class	3.4	4.4	5.7	5.7	4.8
Observed Mismatch	10.4%	10.6%	12.0%	12.9%	13.5%
Part due to the nonmarginal class	8.1	8.3	9.1	9.4	10.3
Part due to the marginal class	2.1	2.3	2.9	3.5	3.2

Source: Clifford C. Clogg, 1980a, "Characterizing the Class Organization of Labor Market Opportunity." *Sociological Methods and Research*, Vol. 8, p. 266.

low income is due exclusively to the marginal class. Consideration of part-time unemployment shows that most of its aggregate level and almost all of its time trend are due to part-time unemployment risks in the marginal class.

The partitioning of observed underemployment into parts due to the latent labor force classes is somewhat analogous to partitioning unemployment into "structural" and "frictional" components. Further, analyses of this kind help to empirically specify theoretical ideas about labor market segmentation. Most labor market segmentation research uses characteristics of occupations or industries to define discrete sectors of the economy with rising or declining labor market opportunities (see Beck, Horan, & Tolbert, 1978; Hodson & Kaufman, 1982; Kaufman, Hodson, & Fligstein, 1983). Our work takes a slightly different approach by defining distinct labor market segments and opportunity structures by matching prior labor market careers to current employment prospects. This emphasis invokes two characteristics that most discussions of labor market segmentation have ignored, namely, marginality as a component of individual careers and marginality as departures from full-time/full-year work at adequate earnings levels. This seems to us a promising avenue for pursuing labor market segmentation research, in addition to using industry and occupational categories (or both, see Petersen, 1994). The rest of this book is devoted to examining the usefulness of our approach.

NOTES

1. Unpaid family workers are an exception to this definition; they are included in the active work force even though their compensation is not easily expressed in a money wage.

2. Levitan and Taggart (1974), among others, provide several techniques whereby the work force is ranked according to "income adequacy" partly satisfying the need for measuring an income from underemployment. However, their procedures represent a major departure from the labor force approach, as can be seen most clearly in their treatment of unemployed workers. The focus of their Earnings Inadequacy Index is to assess the need for income, which is a very different thing from measuring the need for employment. Their index actually ignores some unemployed workers who nevertheless have adequate income (in the form of transfers, for example).

3. Our approach is consistent with many discussions of class in sociology, especially Weberian conceptions that separate discussions of class from relations to the means of production (see Dahrendorf, 1959). We assume that classes in the labor market differ in their opportunity to secure labor market rewards. If the class dimensions of the labor market are measured properly, then our measure of class would explain associations between indicators of labor market rewards.

4. The ILO concept of job productivity, a most difficult characteristic to measure with labor force survey data, is not directly measured in the LUF. However, an indirect measure of work productivity is afforded by the measure of the low-income form just discussed. Since under conditions of perfect competition, the wage income obtained by a worker exactly corresponds to the value of the product produced, the worker who is underemployed by income may also be considered subproductive. Thus, the invisible underemployment measure that we refer to as the low-income form does attempt to satisfy indirectly the need for a *productivity* measure, in the same sense in which income derived from work will indicate the true productivity of work. There is, of course, no exact correspondence between wage income and productivity in a modern economy, and so the low-income underemployed may not represent the subproductive except in an approximate way

5. Using poverty thresholds inflated by reference to the consumer price index and using mismatch criteria for one year—1970—applied to all years

6. Readers are referred to Sullivan (1978) or Clogg (1979, Appendix A) for a more detailed account of each of these measures and a comparison with other measures that have recently been suggested (e.g., Levitan & Taggart, 1974).

7. H_0 tests the hypothesis of mutual independence among the indicators. In this event, the proportions (π) will be determined by simple products of the marginal proportions of the E (employment), I (income), and M (mismatch) variables.

8. Further meaning of this restricted latent structure model can be obtained by noting that the restrictions imply $X = I$ and that Eqs. (2.1)–(2.3) see Appendix 2.4 in this situation imply that this model is equivalent to the conditional independence model which fits the marginals [(EI)(IM)]. This model was actually considered as the hierarchical model M4 presented in Table 2.1.

9. On such a large sample ($n = 59,373$), there is of course no reason to suspect that a two-class model would fit so well.

10. Latent class membership can be predicted from the observed indicators that we have proposed, and the quality of that prediction can serve as another indication of model adequacy. Model H_1, presented earlier, equates the "latent" classes with the observed income classes, and there would of course be no error in predicting "latent" class membership under this model. However, for model H_2 there is uncertainty in predicting latent class membership (see Goodman, 1974b). Under model H_2, 88.8% of the labor force would be correctly assigned into the latent classes by this model (see Clogg, 1980, Table 6, p. 258). The lambda measure of association between X and the joint variable EIM is 0.45, showing that the classes of X can be predicted moderately well by knowledge of the categories of EIM. We would make the most errors in predicting latent class membership for people who have low income but no other kind of underemployment and for people who have low income and are mismatched but working full time. In

spite of the moderately high error rates for these two observed labor force types, the latent class membership of workers can nevertheless be moderately well predicted by H_2. The quality of the prediction of the latent class variable is another criterion whereby various models can be assessed (e.g., in addition to the chi-square statistics measuring closeness of fit of observed and expected frequencies), and in the next section a model that gives a prediction that is better than that of H_2 will be suggested.

Before considering a latent structure model that remedies the substantive defects of the models just discussed, another model will be considered for the sake of completeness. Model H_5 in Table 2.1 tests the hypothesis that all observed underemployment derives from a marginal class and that there exists a single nonmarginal class with zero risk of underemployment. This model is directly related to H_2–H_4 and can be tested by enforcing the restrictions in a two-class latent structure. [Some algebra applied to Eq. (2.3) in appendix 2.4 with these restrictions imposed will prove such an interpretation.] In Table 2.1 we find an $L^2(H_5)$ of 1853.44, showing that such a model is not tenable. This result, when compared with results of H_2–H_4 shows that high risk of mismatch is a characteristic of both the marginal and the nonmarginal latent labor force classes, as further analysis will also show.

APPENDIX 2.1. MEASURING UNDEREMPLOYMENT WITH THE CURRENT POPULATION SURVEY (TAKEN FROM CLOGG, 1979, PP. 213–224)

This appendix details the operational procedures used to measure forms of underemployment. Definitions are made explicit, problems of comparability are mentioned, some alternative strategies are suggested, and the substantive meaning of the new measures is discussed. The material here will be of interest to researchers concerned with measuring underemployment from the Current Population Survey and will hopefully facilitate the constructive modification of our operational procedures. The strategies followed are virtually identical to those used by Sullivan (1978) in her analysis of census data; readers are directed to her study for a fuller exposition of the labor utilization framework and a comparison of it with other approaches (see also Sullivan and Hauser, 1979). It is hoped that with but a few revisions, the procedures presented here can be used by other researchers engaged in the analysis of the labor market and labor force characteristics.

The labor utilization framework was initially conceived as a response to deficiencies in the unemployment rate. A critique of the use of the unemployment rate in developing societies was forcibly developed by Myrdal (1968) and Turnham (1971) among others. That the unemployment rate has also met with severe criticism in the United States is evidenced by the several special committees that have been formed to study its logical makeup and conceptual difficulties. Among these are (a) a 1954 conference held under the auspices of the National Bureau of Economic Research (NBER, 1957), (b) an extensive reassessment of unemployment concepts stemming from the Gordon Committee of 1962 (President's Committee to Appraise Employment and Unemployment Statistics, 1962), and (c) another committee commissioned by President Jimmy Carter, proceedings of which were made available in 1979 (see Sullivan and Hauser, 1979). The International Labour Organization (1976) summarized many of these deficiencies but concluded with the suggestion to measure forms of *under*employment—in addition to unemployment—whether the context is that of a more developed or less developed society. The ILO pointed to the need to measure "invisible" underemployment and singled out the forms of low income, underutilization of skill, and low productivity. Recognizing the extreme difficulty in developing productivity measures on a per-worker basis from standard labor force surveys, Hauser (1974) focused on measures of low income and skill "mismatch" forms, choosing to postpone measurement of the productivity dimension of invisible underemployment until a radically different type of labor force survey could be developed. Many others have proposed frameworks for the measurement of underemployment, and the reader might wish to compare Hauser's approach and our procedures for operationalizing it with them. Of the several frameworks we have reviewed in the course of this work, the approaches summarized by Vietorisz et al. (1975), Gilroy (1975), and Levitan and Taggart (1974) perhaps stand out as plausible alternatives, but the labor utilization framework—or some close relative of it—would provide at least some work force information that these approaches exclude.

Our data are from the March Current Population Survey ("Annual Demographic File") as commercially marketed by DUALABS of Arlington, Virginia. The data are arranged according to hierarchical record types; the raw data contain family records as the first record type, and for each family record, there is one-person record for every nonmilitary person over 14 who is a member of that family. Special computer software is necessary to utilize the information on *both* family and person records. Such programs as CENTS-AID, CENTS-AID version 2, SOS (SPSS override system), TPL, and others will process hierarchical files, the first two at least with admirable efficiency.

To facilitate comparison of our verbal description with actual data codes found in the CPS, we have indicated the data fields within each type of record (family or person) necessary to the determination of the labor utilization variable. An "F" refers to family records, while a "P" refers to person records. For example, "F128, P124" refers to information occurring in the one hundred and twenty-eighth field of family records and the one hundred twenty-fourth field of personal records. All definitions are "standard" definitions used in the CPS and so detailed reference to the official source of definitions is not necessary (see U.S. Department of Commerce, 1963, 1978; Bancroft, 1958; Morton, 1969; Levitan and Taggart, 1974).

Not in the Labor Force

Economic inactivity is said to characterize those persons out of work and not looking for work in the reference week immediately prior to the survey (4, 5, 6, or 7 in P152). This variable is an adjustment for the "major activity" variable (P151), reflecting the usual modifications for fulltime students, housewives, and some others who spend only a small amount of time at work. Those not in the labor force include "students, housewives, retired workers, seasonal workers enumerated in an 'off season who are not looking for work, inmates of institutions, or persons who cannot work because of long-term physical or mental illness or disability."

Included in the not-in-the-labor-force category are the class of persons denoted as "discouraged workers," that is, those persons out of work and not seeking work because they consider job seeking a futile activity. These persons are, in most cases, a "ready reserve" labor supply, since with an increase in demand they would presumably begin seeking work and so are not disassociated from the labor market as the labor force concept makes them appear. Considerable fluctuation in their numbers occurs with economic growth or recession. Consideration of discouraged workers could alter demographic comparisons based upon unemployment rates or upon labor force participation rates and would accordingly distort perceptions of labor market change over time.

Willard Wirtz, secretary of labor under President Lyndon B. Johnson, considered discouraged workers in his indexes of underemployment, and this was to some extent responsible for his findings of 45% underemployment (or "subemployment") in urban ghettoes. This discouraged worker presents problems not only in the comparison of rates but also in attempts to model the working life of the individual (e.g., as in tables of working life) since it is unclear how to calculate base populations "exposed to the rise of employment," given the borderline status of these marginal labor force participants.

Note also that the not-in-the-labor-force category does not include persons such as new *entrants* and *reentrants* to the labor force, whose labor force "participation" is nearly as tenuous as that of the discouraged worker. Sullivan, using 1970 Census data, was able to isolate entrants and reentrants for special analysis; however, we have been unable to locate them for a similar analysis with the March CPS data.

Subunemployed

Ideally, for reasons mentioned above, we would like to isolate discouraged workers for the first category of underemployment, following that approach used by Wirtz and subsequently used in the "subemployment" index or in Levitan and Taggart's (1973) Index of Earnings Inadequacy. However, questions aimed at determining the volume of discouraged workers (e.g., "What is the reason for not looking for work in the past month?") are not part of the March series for the CPS. But a proxy for them can nevertheless be obtained. If a person not in the labor force during the reference period of the survey (4, 5, 6, or 7 in P152) was a *part-year* worker whose reason for part-year work was that he or she was "looking" for full-year work (1 in P146), then our claim is that the given individual resembles a discouraged worker. No other questions in the March CPS, to our knowledge, allow a clearer identification of discouraged workers. Because our method of locating these members of the marginal labor force is a crude one, we prefer to designate the category thus contained as the "sub-unemployed." Because of the confusion over the tenuous work force status of these types of persons, we presented in Chapter 2 various ratios and rates reflecting different base populations.

The labor force consists of those persons working, persons with jobs but not at work, or persons looking for work (1, 2, or 3 in P152). All of the subsequent definitions pertain to members of the standard labor force.

Unemployed

Unemployed persons "are those civilians who, during the survey week, had no employment but were available for work and (a) had engaged in any specific job-seeking activity within the last four weeks" (3 in P152) or (b) were with a job but not at work and were "waiting to be called back to a job from which they had been laid off . . . or . . . were waiting to report to a new wage or salary job within 30 days" (2 in P152 and 5, 6, or 7 in P159).

All persons aged 14 and over who were at work (1 in P152) or were "with a job but not at work" but were temporarily absent because of illness, vacation, bad weather, labor dispute, or other (noneconomic personal) reasons (2 in P152, 1, 2, 3, 4, or 8 in P159) are the employed. With the 1970 Census and in several Bureau of Labor Statistics Publications prior to that time, the formal working age on which unemployment statistics are based was raised from 14 to 16 years. We maintained the 14-year minimum in some of our work; distortions to crude rates by reason of the inclusion of ages 14 and 15 can be expected to be minimal. These formally employed persons, with some modifications, are subject to the subsequent sorts.

Underemployed by Low Hours

This category is equivalent to the "part time unemployed" or the "part time for economic reasons" category already in wide use. The Census Bureau and the Bureau of Labor Statistics use 35 hours as the minimum fulltime workweek. Any person working 34 hours or less is subject to the sort for "under-employment by hours," with the following qualifications.

Only employed persons who are "at work" during the reference week are classified by hours of work (l in P152). Employed persons not at work during the reference week (2 in P1521 and 1, 2, 3, 4, and 8 in P159) are assumed to be adequately utilized by hours of work. Although this is certainly not the optimal strategy for determining involuntary part-time unemployment, this expedient is necessary because no questions are asked of employed persons not at work regarding their hours of work in the week last worked. To the extent to which employed persons not at work (by virtue of personal reasons, etc.) are in fact working less than the full workweek, the category of underemployed by hours can be expected to provide a *conservative* estimate of the real extent of part-time employment. The number of workers not working but counted as employed is moderate and so could affect our estimates by as much as .2%. The wish to conform to a definite time referent, useful in computing rates and gauging trends over time, is responsible for the CPS format regarding these persons. We have accordingly determined the part-time unemployed (i.e. the part-time employed for economic reasons) following this format. Part-time workers are defined as those working less than 35 hours during the survey week (1–34 in P153–154). Economic reasons for working less than 35 hours include "slack work, material shortages, repairs to plant or equipment, start or termination of job during the week, and inability to find full-time work" (1, 2, 3, 4, 5, or 6 in P157–158).

A caveat is necessary concerning the extent to which short hours "for economic reasons" is a measure of diminished demand for labor. The reason for part-time work of material shortage is not necessarily a reflection of demand for labor but is instead a measure of the extent to which the material factors of production are in short supply. The reasons listed as "repairs to plant or equipment" and "start or termination of job during the week" are also not necessarily measures of slack demand for labor. These reasons are perhaps gauges of *frictional* part-time employment. Only the reasons "slack work" and "inability to find full-time work" are true measures of the extent to which the economy is unable to provide a full workweek to those who demand such work. We can probably surmise, however, that the variation in the underemployed-by-hours category over time will result principally from reasons reflecting the variable demand for labor. In dealing with the underemployed-by-hours category, then, we encounter a difficulty in interpreting our operationalization of Hauser's LUF as a measure of slack demand for labor. Future applications of this framework could take these criticisms into account; our wish to conform to standard definitions insofar as possible has resulted in this deficiency.

Note that with our modifications thus far, the category of unemployment is equivalent to the standard definition of unemployment, and the category of underemployed by hours is equivalent to the BLS's category of "part time for economic reasons." These two categories are *jointly* compatible with the similar BLS categories that have been used for many years.

Underemployment by Low Income

With the underemployed-by-low-income category, we make a minor departure from the approach of Sullivan. The comparability of our results with Sullivan's will thus be smallest when considering this form. The difference pertains to the determination of the proper income thresholds; this critical issue is far from resolved by Sullivan's or our own approach, even if it is agreed that the poverty thresholds of the Social Security Administration should be somehow used as the basis for this determination. The problem of defining underutilization by low income is critically dependent upon the distinction between family versus individual and primary versus secondary earners. All approaches thus far fall short of the desideratum because of the extreme difficulty in determining the influence of the decision processes within households with two or more earnings that concern the allocation of employment potential to the labor market. The degree of involuntariness of underemployment by income (as well as by other criteria) is a necessary datum to unambiguous classification of workers within the same household but unfortunately cannot be ascertained in the usual labor force survey. Conceivably, employment producing a low wage rate is much more "severe" for heads of households or for primary earners than it is for nonheads or secondary earners.

Persons previously classified are not subject to our income sort. In addition, voluntary part-time workers are not subject to the income sort because they are assumed to earn low wages in proportion to their voluntarily shortened hours. Full-time students (i.e. persons whose major activity was "school") were excluded from the income sort for reasons similar to those given by Sullivan (1978). Therefore, with the exception of voluntary part-time workers, full-time students, and persons previously classified, the universe to which the income sort applies are full-time workers (1 in P152 and P152–154 greater than 34). This universe also contains those persons "with a job but not at work" (2 in P152 and 1, 2, 3, 4, or 8 in P159).

The universe on which Sullivan's income sort applied also omitted "new entrants" and "reentrants" to the labor force. New entrants are "persons who never worked at a full time job lasting two weeks or longer." Reentrants are "persons who previously worked at a full-time job lasting two weeks or longer but who were out of the labor force prior to the beginning to look for work." Entrants and reentrants to the labor force may be either employed or unemployed. The CPS data only provide information on entrants and reentrants *who are unemployed* and, further, only on those unemployed persons who are looking for work (P 168). These unemployed entrants and reentrants are, of course, already classified as unemployed. Therefore, in our use of CPS data, entrants and reentrants who are working cannot be deleted from the universe to which the income sort applies, which is different from Sullivan's analysis of Census data. Entrants and reentrants who are unemployed are already classified as unemployed. Entrant and reentrants who are employed (i.e. those who have recently entered the labor force and found work) may have little or no work-related income for the *preceding* year, complicating the determination of the low-income form for them.

With the universe selected according to this procedure, it was necessary to select income thresholds that could serve to define underemployment by low income. The determination of income thresholds for each worker within a household is based upon the poverty thresholds derived by the Social Security Administration and used in the CPS. No other set of income levels, to our knowledge, has been developed with the underlying criterion of minimum subsistence in mind. (Many other informal measures of "poverty" are, of course, used, but these are largely based on ad hoc conceptions of poverty. Clearly one could adjust poverty thresholds and find as much poverty as desired.) The Social Security Administration's index gives a figure for total family income that is a minimum annual household expenditure necessary to ensure at least subsistence levels of household welfare. These thresholds are based on age and sex of the household head, size of family, and farm/nonfarm residence of the family. Since the income data on the CPS are based upon the preceding year, a departure from the desired time-reference comparability is encountered here. The preceding categories, of course, did have the desired time comparability with the usual unemployment statistics.

Briefly, the income sort, given the universe above, consists of the following steps;

1. *Work related income* is calculated for each worker. This is the sum of (a) wage or salary income, (b) nonfarm self-employment income, and (c) farm self-employment income (P67–72, P73–78, P79–84) Work-related income excludes unearned income such as Social Security, dividends, interest, rental income, welfare and unemployment compensation. The assumption is that work-related income for the preceding year can be used as a proxy for income for the current year. This assumption implies that the income category of the LUF will not be particularly sensitive to short-term fluctuations in the demand for labor. (However, as we indicated in Chapter 5, this measure of low-income form has certain advantages, e.g. as an indicator of marginality that can be used partially to explain the occurrence of the other forms.)

2. Annual work-related income is converted to an *equivalent weekly wage*, using the "weeks worked last year" variable (P131).

3. Poverty thresholds for every worker in the household were determined. For the chief income recipient (not necessarily the head of the house), the cutoff point was assumed to be the threshold for the family as a whole. These cutoffs vary with age and sex of household head, size of family, and farm/ nonfarm residence.

4. For secondary earners, we used the poverty threshold for a primary individual with the same age, sex, and residence characteristics. Sullivan applied the poverty thresholds far a single urban (i.e. nonfarm) male under 65 years of age uniformly to all secondary male workers, and she applied the poverty threshold for a single urban (i.e. nonfarm) female under 65 years of age uniformly to all secondary female earners. We include the age and residence detail, in addition to sex detail, for the secondary workers so that the criteria for partitioning secondary workers be made similar to that used for primary workers. This constitutes a slight modification of Sullivan's approach.

5. For members of subfamilies (4 in F27–29 or 4 in P27–28), an alternative strategy had to be employed altogether, since poverty status is not determined in the CPS for subfamilies. The reason for this curious omission is that subfamilies are considered to be dependent on some primary family, the poverty status of which has already been determined, assuming that the subfamily is dependent on the primary family. Rather than adjusting income levels for primary families containing subfamilies and then treating both as separate families on which the original criteria for determining thresholds would be used, we chose to treat workers in subfamilies as if they were primary individuals. We are merely assuming that workers in subfamilies, by virtue of their formal dependence on a primary family, should be regarded in the same way as we regard secondary workers in general. Weekly income thresholds are presented in Table A. 1, pg. 222 of Clogg, 1979.

Underemployed by Mismatch

All persons not included in the above categories are subject to the mismatch sort, including voluntary part-time workers. The mismatch category is the most difficult to interpret, owing partly to the artificial means by which "overeducation" is inferred. We apply directly the cutoff points for completed years of education for educationally heterogeneous occupational strata used by Sullivan. For years 1969–1970, the CPS used the 1960 Census occupational classification. For these years, Sullivan's adjustments for the 1960 Census were utilized. For years beyond 1970, we use her education cutoffs for the 1970 Census occupational classification. For the details of her procedures, including adjustments for occupational strata dominated by females (whose educational attainment is typically higher than that of males for some occupational groups) the readers is referred directly to her work (1978). Completed years of education are obtained from P57–58 and P59. Current occupation, necessary for determining mismatch in relation to current economic conditions, is obtained from the detailed code of P172–174, the three-digit occupational codes developed by the Census Bureau.

The rationale used in the mismatch classification is essentially the following. Workers are denoted as mismatched if their own completed years of education is more than one standard deviation above the mean completed years of education for their occupational group. Such an operational strategy is clearly open to charges of being arbitrary and somewhat artificial, and for this reason, absolute levels of mismatch are extremely difficult to interpret by themselves. However, we have directly applied the educational cutoffs determined for 1970 uniformly to all years of the study so that across-time change in mismatch levels can be inferred. Changes in real mismatch, owing to diminished or increased demand for skilled labor, are approximately measured by our time-series measures of mismatch. The principal deficiency of our approach is that the shape (including the skew) of educational distributions within occupational strata might change over time in ways unrelated to the extent of true mismatch. Educational requirements for entry to an occupation might increase, perhaps reflected in higher mean education, giving the impression that younger workers are mismatched relative to older workers. Or, with technological innovation, the educational requirements of the specific occupations within an occupational strata might change over time, resulting in both a change in location and a change in scale, again overstating the extent of mismatch. Other factors can also conceivably distort the mismatch measure. These criticisms are mitigated somewhat in this study by the fact that only a short time span (5 years) is considered, and the changing shape of the educational distribution in the short run is probably mainly the result of real change in mismatch. In determining mismatch trends over the long run, perhaps some type of annual adjustment would have to be devised.

Although the mismatch category is perhaps the least satisfactory measure of underemployment in this study, we are not aware of a more rational measure to determine it than Hauser's original suggestion.

Adequately Utilized

Adequately utilized workers are merely the residual category left over after the above sorts have occurred. No attempt is made in our work to separate this category into full- and part-time adequately utilized, as was done by Sullivan.

APPENDIX 2.2. ADDITIONAL ANALYSES

To examine model H_2 in more detail, a series of other models closely related to it (H_3, H_4, and H_5) were considered. Table 2.1 also contains the fit of these models, and they are now discussed in turn.

In Table 2.3, the estimated risk of part-time employment is 0.2% for the nonmarginal class, and therefore it is natural to examine the statistical contribution of this nonzero estimate to the fit of the model. Model H_3 tests the hypothesis that all observed unemployment and part-time employment derives from the marginal class and is obtained from H_2 by applying the restriction that $\pi_{22}^{EX} = 0$ (in addition to the restriction that $\pi_{12}^{EX} = 0$). With $L^2(H_3) = 6.77$, this model is clearly acceptable on purely statistical grounds. Model H_4 tests the hypothesis that all observed underemployment except mismatch derives from the marginal class and imposes on H_3 the additional restriction that $\pi_{12}^{IX} = 0$. With $L^2(H_4) = 10.28$, this model is also acceptable on purely statistical grounds. However, models H_2, H_3, and H_4, which all stem from the unrestricted two-class model and which all fit the data well, should be rejected on substantive grounds. All of these models either assume that the risk of unemployment or part-time unemployment is nil or estimate that risk is negligible. Because the "frictional" kind of unemployment and part-time employment occurs to both marginal and nonmarginal workers, implying that estimates of the risk of these forms of underemployment should be nonnegligible for the nonmarginal class, these models are unacceptable. Some frictional or temporary unemployment, for example, will necessarily occur to nonmarginal workers as they progress through the chain of jobs that constitute their career life cycle, and therefore any model that determines a zero risk of unemployment for nonmarginal workers should be rejected.

APPENDIX 2.3. MODEL DERIVATION

Consider a six-class latent structure where

$$\pi_{12}^{\bar{E}X} = 1; \quad \pi_{1t}^{\bar{E}X} = 0, t = 3, 6$$

$$\pi_{23}^{\bar{E}X} = 1; \quad \pi_{2t}^{\bar{E}X} = 0, t = 2,4,5,6$$

$$\pi_{14}^{\bar{I}X} = 1; \quad \pi_{1t}^{\bar{I}X} = 0, t = 2,3,5,6$$

$$\pi_{15}^{\bar{M}X} = 1; \quad \pi_{1t}^{\bar{M}X} = 0, t = 2,3,4,6$$

$$\pi_{36}^{\bar{E}X} = \pi_{26}^{\bar{I}X} = \pi_{26}^{\bar{M}X} = 1$$

It may be verified on algebraic analysis of Eq. 2.3 with the above restrictions that, for the cells denoting any type of multiple underemployment, the proportion in each cell derives exclusively from a latent marginal class (where $t = 1$). It may also be verified that, for the cells denoting absence of multiple underemployment, the following relationships hold:

$$\pi_{122} = \pi_2^X + \pi_1^X \pi_{11}^{\bar{E}X} \pi_{21}^{\bar{I}X} \pi_{21}^{\bar{M}X} \text{ (unemployed only)}$$

$$\pi_{222} = \pi_3^X + \pi_1^X \pi_{21}^{\bar{E}X} \pi_{21}^{\bar{I}X} \pi_{21}^{\bar{M}X} \text{ (part-time unemployed only)}$$

$$\pi_{312} = \pi_4^X + \pi_1^X \pi_{31}^{\bar{E}X} \pi_{11}^{\bar{I}X} \pi_{21}^{\bar{M}X} \text{ (low-income only)}$$

$$\pi_{321} = \pi_5^X + \pi_1^X \pi_{31}^{\bar{E}X} \pi_{21}^{\bar{I}X} \pi_{11}^{\bar{M}X} \text{ (mismatch only)}$$

$$\pi_{322} = \pi_6^X + \pi_1^X \pi_{31}^{\bar{E}X} \pi_{21}^{\bar{I}X} \pi_{21}^{\bar{M}X} \text{ (no underemployment)}$$

For each of these cells, the manifest proportion (e.g., π_{122}, denoting unemployment only) is composed of a part due to a nonmarginal class (namely, π_2^X) and a part due to the marginal class (namely, $\pi_2^X \pi_{12}^{\bar{E}X} \pi_{21}^{\bar{I}X} \pi_{21}^{\bar{M}X}$). This six-class latent structure can be thought of as a special kind of two-class model where now for a single nonmarginal class the risk parameters are redefined as $\pi_i^X / (1 - \pi_1^X)$ for $i = 2, \ldots, 6$. For example, $\pi_2^X / (1 - \pi_1^X)$ may be taken as the risk of unemployment for a nonmarginal class, the quantity $\pi_3^X / (1 - \pi_1^X)$ may be taken as the risk of part-time unemployment for a nonmarginal class, and so forth. Goodman (1975) considered a model very similar to the present one by choosing certain "scale types" (e.g., response patterns corresponding to a

Guttman scale). We see that his model, on the face a six-class latent structure model [with very different "scale types"] can be thought of in this context as a reasonably simple kind of two-class model. Multiple underemployment is prohibited in a single latent nonmarginal class here, and that is all that distinguishes this model from the unrestricted two-class model H_2 presented earlier. Conditional independence governs the relationship among the observed variables in the marginal class, but conditional quasi-independence governs that relationship in the nonmarginal class. In certain respects this is a simpler interpretation of Goodman's new model for the scaling of response patterns, because we need consider only two (and not six) latent types of individuals.

APPENDIX 2.4. THE LATENT STRUCTURE MODEL

Let π_1^X, π_2^X, ..., π_T^X denote the proportions of the labor force in the T latent labor force classes. (The number of classes—T—is left unspecified for now, but later we show that only two classes are necessary to explain these data.) Let π_{it}^{EX} denote the conditional probability that a member of the t-th latent class will be in the i-th category of E. The quantity π_{1t}^{EX} is the risk of unemployment for a member of the tth class, and the quantity π_{2t}^{EX} is the risk of part-time employment for a member of the t-th latent class. The quantity $\pi_{3t}^{EX} = 1 - \pi_{1t}^{EX} - \pi_{2t}^{EX}$ is the risk of being employed full time. Parameters π_{1t}^{IX} and π_{1t}^{MX} are defined similarly, denoting the risk or probability of low income and mismatch, respectively, for members of the t-th latent class. These risk parameters enable us to characterize labor force class structure in the framework of the latent structure technique; they are merely "rates" of unemployment, part-time unemployment, low-income underemployment, and mismatch for each of the T latent labor force classes.

If the T latent classes were observable, then the UH × I × M table could be replaced by an observable 3 × 2 × 2 × T table where the last dimension of this table would now refer to the labor force class variable. Letting $\pi_{ijkt}^{E,I,M,X}$ represent the proportion in the (i,j,k,t) cell of this hypothetical table, and assuming that the indicators of labor market rewards are conditionally independent of one another given the labor force class t, application of the law of conditional probability produces

$$\pi_{ijkt}^{E,I,M,X} = \pi_t^X \pi_{it}^{\bar{E}X} \pi_{jt}^{\bar{I}X} \pi_{kt}^{\bar{M}X} \tag{2.1}$$

However, the UH x I x M table does not contain estimates of $\pi_{ijkt}^{E,I,M,X}$ but rather estimates of

$$\pi_{ijk} = \sum_{t=1}^{T} \pi_{ijkt}^{E,I,M,X} \qquad (2.2)$$

or marginals over the unobserved classes of the latent variable X. The latent variable X, which is interpreted here as the underlying mechanism dividing the labor force into latent classes, 1, the latent classes are confounded or mixed and estimates of the aggregate risk to a particular form of underemployment obtained from the observed table will not estimate the true risk characterizing each "class," except in the special case where there is only one "class." By substituting the right-hand side of Eq. (2.1) into Eq. (2.2), the latent structure model expresses the manifest proportions π_{ijk} in terms of risk parameters according to the following equation:

$$\pi_{ijk} = \sum_{t=1}^{T} \pi_t^{X} \pi_{it}^{\bar{E}X} \pi_{jt}^{\bar{I}X} \pi_{kt}^{\bar{M}X} \qquad (2.3)$$

(see Lazarsfeld & Henry, 1968; Goodman, 1974a,b). Equation (2.3) says that: (1) there are T mutually exclusive and exhaustive latent classes in the labor force, and (b) within each latent class the indicators E, I, and M are mutually independent. Readers unfamiliar with the latent structure technique can gain understanding of Eqs. (1)–(3) by drawing the analogy to factor analysis. The latent variable X in these equations is assumed to account for the association among indicators, just as in a factor analysis an underlying factor is assumed to account for correlations among indicators. We are in effect assuming that E, I, and M can serve as indicators of X and are using a statistical technique appropriate for the case where both the indicators and the latent variable are discrete entities. Bringing the labor force measures I and M into consideration enables us to characterize the latent class structure X to the labor force by identifying risk parameters of the kind $\pi_{it}^{\bar{E}X}$ for each distinct labor force class. Without this additional information, such risk parameters could not be identified (see Goodman, 1974a). With only the information in the time form of underemployment contained in the E variable, labor force class structure would have to be characterized by methods different from the ones used here.

3

Analyzing Trends
in Labor Force Activity

INTRODUCTION

We have spent the last two chapters dealing with basic issues in the definition of labor force activity and with different ways of measuring it. We now turn our attention to basic analyses of trends in labor market activity using the measures and conceptual apparatus we developed in the last chapter. We hope that by the end of this chapter readers will understand a number of things about labor force participation trends through the 1970s. Foremost among these is that there are systematic differences in the labor force activity of women and minorities in the U.S. labor market. These differences are detectable using the methods we outlined and these methods require relatively little in terms of ambitious data collection or expensive measurement development.

This chapter discusses findings of an analysis of changing labor force composition and underemployment trends (Clogg & Sullivan, 1983), a cohort analysis of recent trends in labor force participation (Clogg, 1982a) and the question of whether systematic changes in the demographic composition of the labor force are associated with trends in unemployment and underemployment (Clogg & Shockey, 1985).

As we have tried to emphasize, studying trends in labor market activity is not merely an academic question. Trends in underemployment across demographic groups tell us a lot about how labor markets work and about the overall integration of traditionally disadvantaged groups into the economic and social mainstream of society. These analyses also inform debates about the

overall health of the labor market, rising (or declining) social inequality, social
alienation, debates about job skills and the match between job opportunities
and workers' skills.

LABOR FORCE COMPOSITION AND
UNDEREMPLOYMENT TRENDS, 1969–1980

The work by Clogg and Sullivan (1983) discusses the conceptual and normative reasons for developing indicators of underemployment. The measurement and refinement of measures of underemployment took on an increased
urgency during the recession of the early 1980s. Clogg and Sullivan's work
developed in the context of near record unemployment rates. These high unemployment rates were a threat to those still employed because jobs could be
downgraded during recessions. Potential employees in the labor market would
be forced to "settle for less" because the labor market provided few alternatives to downgraded jobs. Conventional measures of labor force activity such
as the labor force participation rate and the unemployment rate ignored marginal employment as a social and economic problem.

The lack of satisfying underemployment indicators did not reflect a lack
of interest on the part of researchers or policymakers. Researchers have recognized the detrimental effect of inferior employment on individual and social welfare at least since the 1930s (see Robinson, 1936). Researchers agree
that underemployment manifests itself in a small set of symptoms (ILO, 1957,
1966, 1976; a detailed historical treatment appears in Sullivan, 1978, pp. 3–13
and 24–44). These symptoms include involuntary part-time work, "working
poverty," the gross mismatch between educational credentials and employment opportunities, and discouraged workers who end the active search for
work.

Despite widespread adoption of the term *underemployment* and a con-
gressional mandate to collect systematic data [Pub. L No. 93-203, 1973, S.302
(b), (c) and Pub L. 94-444, 1986, S.13 (d)], research on underemployment indicators languished for much of the 1970s. Aside from political reasons for
postponing their development, researchers were stymied by two technical issues. First, observers disagreed about the proper way to measure the symptoms of underemployment. Second, some researchers wrongly concluded that
it was necessary to use a composite underemployment index because techniques for the adequate analysis of multistate indicators were becoming available (e.g., NCEUS, 1979, pp. 77–79).

In this section we summarize an analysis of labor market trends from
1969 to 1980 using the LUF. We then present an analysis of the effects of the
demographic composition of the labor market on underemployment. The analy-

sis does a good job of highlighting the utility of the LUF as an objective social indicator and a normative social indicator suitable for policy analyses of labor market opportunity.

The Record of Annual Changes in U.S. Underemployment, 1969–1980

We described earlier the definition of LUF components used in our analysis: (1) *subunemployment* (a proxy for discouraged workers), (2) *unemployment*, (3) *involuntary part-time work*, (4) *full-time workers* whose work-related income is below some specified poverty threshold, (5) *mismatch*, which measures gross overeducation relative to that possessed by others in similar occupations, and (6) *adequate employment*. The poverty threshold is defined as 125% of the individual poverty threshold as defined by the Social Security Administration. The mismatch category is defined using 1970 as the benchmark.

Figure 3.1 presents the distribution of LUF components as a percentage of the labor force. The proportion of the economically active population with adequate employment (not shown in Figure 3.1) declined from 77.0% in 1969 to 67.4% in 1980, a decline of almost 10%.

The increased relevance of mismatch is the most salient trend in Fig. 3.1. Educational mismatch increased from 7.8% in 1969 to 14.2% of 1980. Expressed in different terms, this is an 80% increase in mismatch, or an average increase of 0.6% per year. Because the average size of the labor force from 1969 to 1980 was 100 million, these figures correspond to an increase of over 6 million mismatched workers over the 12-year period.

In spite of our crude measurement, it seems clear that there has been a sustained and dramatic increase in the prevalence of mismatch. This finding

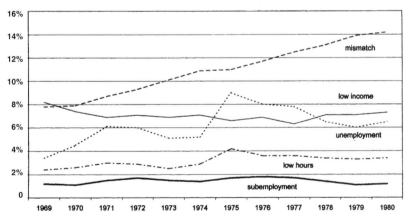

Figure 3.1. The proportion of adequately employed declines; the mismatch between education and job opportunities increases. Source: Clogg & Sullivan, 1983, p. 129, Table I.

supports Freeman's (1976) conclusion that there has been overeducation relative to the demand for educated labor. Our result is inconsistent with the alternative explanation by Featherman and Hauser (1978) that there is only apparent decline in the occupational return to college education.[1]

Contrary to the charges of critics (see Keyfitz, 1968) mismatch does not have to rise as educational attainment rises. Mismatch only rises if the occupational structure does not accommodate increases in education. The mismatch component of the LUF scheme is more suited to policy and programmatic discussion of the overeducation phenomenon than are occupational status or income attainment models because they evaluate educational fit across the full range of the occupational distribution rather than focusing only on the most educated workers. Further, the LUF leads directly to a measure of prevalence, not a regression coefficient usually calculated on a single cross-sectional or retrospective survey of occupational origins and destinations.

While there is a strong secular trend in occupational mismatch, other LUF components are closely attuned to the business cycle. Our measures of subunemployment, unemployment, and involuntary part-time employment change in recession years (e.g., 1975, 1976). The low-income component appears to stay close to 7% over the whole range of the study.[2]

The data in Fig. 3.1 might be condensed in several ways without serious loss of information. One approach would be to combine subunemployment, unemployment and part-time employment components into a single "economic underemployment" category, retaining the mismatch component as a separate category. When we do this with the data in Fig. 3.1 full employment (1969) is associated with 15.2% economic underemployment and 7.8% mismatch; the recession year 1975 is associated with 21.5% economic underemployment and 11.0% mismatch; and the last year (1980) records 18.4% economic underemployment. The difference between recession and full employment for the periods considered here is 5.9% or 6 million workers who shift downward during recessions.

The picture of labor force structure and trend, and the implications for a programmatic response, are radically different when the data of Fig. 3.1 are used in place of the single measures of unemployment alone. Specific policy proposals would shift from reducing unemployment during recessions to the creation of high-quality jobs to match workers' educational achievements and structuring incentives to create full-time jobs for part-time workers who want them.

Trends by Gender

Underemployment varies considerably by sex, age, and race, as well as by other factors, and a central preoccupation of labor market researchers is to study the labor market prospects of demographic groups.

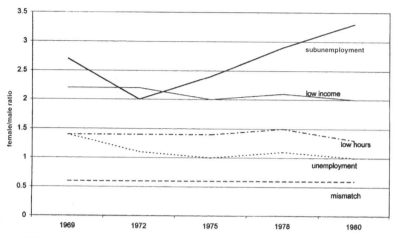

Figure 3.2. Relative to men, women in the active labor force have stayed in place or slightly improved their prospects; those outside the labor market face greater difficulties. Source: Clogg & Sullivan, 1983, p. 129, Table II.

Figure 3.2 presents the LUF components by sex for selected years of the study. Even with dramatic changes in the gender composition of the labor force, the ratio of female to male prevalence rates for most of the LUF components remains very stable. The conventional labor force participation rate appears to underestimate labor supply. Low income employment, involuntary part-time employment, and unemployment all declined slightly for women during this period, and educational mismatch remained stable at 60% of the rate for men. In fact, the figure suggests that there is a growing separation between women in the active labor force whose labor market prospects remain stable or improve slightly and growing numbers of women outside the active labor force who are outside of the sphere of expanding opportunity. This is especially obvious when looking at trends in discouraged workers. The most substantively interesting result is the change in the subunemployment component where female/male ratios went from 2.0 in 1972 to 3.3 in 1980. Clearly, women are increasingly involved and marginalized in the labor market.[3]

Trends by Age

Figure 3.3 presents LUF components by age for selected years. One criticism of the LUF is that mismatch is only a pertinent concept for the youthful members of the labor force. In 1969 mismatch is indeed the highest for 20- to 34-year-olds (9.1%), but it is nearly as high (7.1%) for 50- to 64-year-olds. By 1980 mismatch had increased 195% for 20- to 34-year-olds (to 17.7%), but also increased 170% for the oldest primary working age group (to 12.1% for 50- to

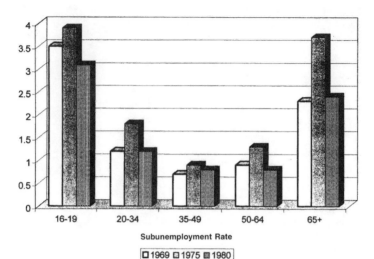

Figure 3.3. LUF categories by age, selected years.
Source: Clogg & Sullivan, 1983, p. 132, Table III.

64-year-olds). Using the *same* definition of mismatch for 1969 as for 1980, it is evident that the 50–64 age group was more mismatched in 1980 than the 20–34 age group in 1969.[4] Although previous studies have indicated substantial downward occupational mobility among some older workers (Leigh, 1978, p. 81), the mismatch indicator provides a less cumbersome and more systematic record of the trend.

The age differentials in the other LUF components show how the rewards

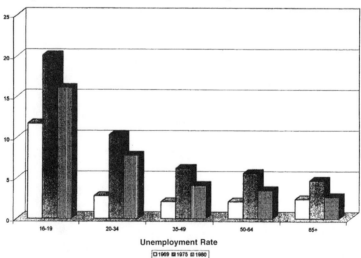

Figure 3.3. Continued.

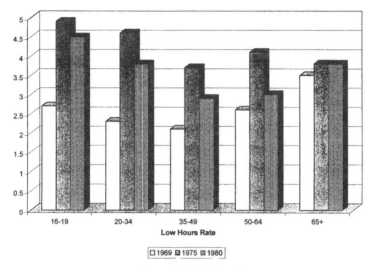

Figure 3.3. Continued.

of the labor market are distributed by age, or how underemployment varies in a U-shaped manner for all of the components, except mismatch. The very young (16–19) bear the greatest burden with respect to subunemployment and underemployment; they are similar to workers 65 and over with respect to part-time work. The low-income measure is very high among youth and even higher among the oldest workers. The age group 20–34, in addition to having the highest mismatch rates, also has very high rates of unemployment,

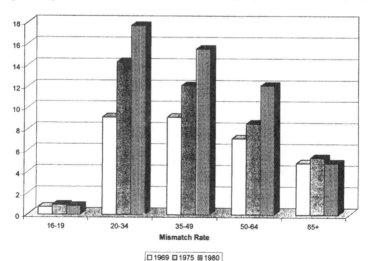

Figure 3.3. Continued.

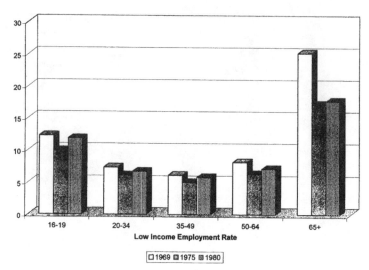

Figure 3.3. Continued.

part-time unemployment, and subunemployment. In short, the 20–34 age group has the most difficulty acquiring the major rewards of the labor market—a full- or part-time job, adequate income, and use of their skills. Around 37% of those in this age group are underemployed in both 1975 and 1980.

Trends by Race

Figure 3.4 presents the LUF components by race, comparing blacks with nonblacks. The available labor force indicators in 1980 yielded two generalizations about blacks (e.g., see U.S. Bureau of Census; 1980b, pp. 60–61). First, black labor force participation rates are close to nonblacks, although there has been a persistent decline in the participation rates of black males. Second, since World War II there has been a nearly constant two-to-one ratio between black unemployment rates and the unemployment rates for nonblacks. Figure 3.4 suggests that there is more to the story than this when we look at the expanded information provided by the LUF.

Figure 3.4 confirms that black unemployment remains double that of nonblacks for most of the 12-year period. However, there is a noticeable jump in the black/nonblack ratio for subunemployment as blacks in the discouraged worker category increase at a greater rate than others. The growth of discouraged workers may help explain declining participation rates, especially among black men.

On the other hand, there is a modest decline in the black/nonblack ratio

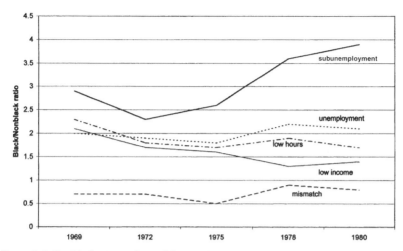

Figure 3.4. For blacks, inequality in labor market opportunities increase over the decade: Low hours and low-income work decline; unemployment and discouraged workers rise. Source: Clogg & Sullivan, 1983, p. 132, Table IV.

for the hours worked and low-income measures. Mismatch shows some trend toward "parity" between blacks and nonblacks, indicating greater educational attainment for blacks and perhaps also greater difficulties in receiving occupational rewards commensurate with their educational attainment. The adequate employment ratios remain steady over time, but inspection of the various forms of underemployment shows that offsetting trends have canceled each other out.

Overall, these results support the conclusions of Wilson (1987, 1996) and others regarding the changing labor market prospects of blacks. Blacks are much more likely to be among the ranks of discouraged workers and their representation in this category has risen. However, blacks who manage to avoid unemployment are improving their position relative to the rest of the labor market. These results provide modest support for research suggesting that blacks are dividing into an upwardly mobile middle class and an economically disadvantaged class left behind in the wake of expanding opportunities.

THE RELATIVE EFFECTS OF DEMOGRAPHIC AND OTHER TEMPORAL CHANGES ON AGGREGATE UNDEREMPLOYMENT

This section applies standard log-linear models (Goodman, 1971; Haberman, 1978) to a contingency table that is basic to analyses of trends in labor force activity (see Duncan, 1975, 1981; and Clogg, 1979, pp.45–76, and Appendix 1, Chapter 1).

The Basic Hypothesis

Some analysts attribute several of the percentage points in the unemployment rate to the increased participation of demographic groups that usually have high unemployment (Perry, 1970; Wachter, 1976; Flaim, 1979). Alternatively, underemployment may increase because the rates of underemployment have increased, either for the entire labor force or for significant sections of it. Here we investigate the hypothesis that changes in the demographic composition of the labor force contribute to the secular rise in underemployment.

We consider five variables: Age (16–19, 20–34, 35–49, 50–64, 65+), Gender, Race (black, nonblack), LUF (six categories), and Time (years 1969–1980). Examining the net effect of age on LUF is the first step toward specifying the life cycle dynamic associated with labor market activity (see Winsborough, 1978). Gender differences in underemployment are important in an era of concern about women's relative success in securing labor market rewards. Race differences in labor market experiences are important to policy analysts. Incorporating time into the analysis allows us to separate the effects of the changing composition of the work force from the effects of the business cycle and other time trend effects.

The methods for fitting log-linear models to these data are described in Appendix 3.1. The results support the hypothesis that there is substantial contribution to underemployment effects embodied in the changing demographics of the labor market. The composition of the labor force accounts for between 29 and 31% of the total period effect in underemployment. Another 18% of the total period effect is explained by changes in underemployment rates that are tied to specific groups.

Results for Specific Demographic Groups

Results across specific groups suggest that *differences by age are by far the most important compositional change in the labor force from 1969 to 1980*. Age accounts for 79% of the total demographic compositional change in the labor force. Gender is also substantively important, involving some 17% of the total period effect. By contrast, race and other more complex combinations of demographic groups' changes account for only about 4% of the compositional shift in the labor force. Using standardization techniques like those in Clogg (1979, pp. 45–76) suggests that much of the secular rise in underemployment from 1969 to 1980 can be attributed to the rise in the proportion of labor market participants in the 20–34 age group.

Having analyzed the influence of period on the various types of associations that can be of interest in the Age (A) by Sex (S) by Race (R) by LUF (L) by Time (T) contingency, we now consider a tentative model for the data. Model building strategies described in Clogg (1979, pp. 45–76) were used to specify potential models, and standard procedures (Bishop, Fienberg, & Holland, 1975) were used to determine the adequacy of the results. The hierarchical log-linear model that was settled on fits the marginals (ASRL), (ASRT), (ASLT), and (RLT). The first (four-way) marginal in this configuration corresponds to the "structural constraint" that underemployment varies by age–sex–race group. The second (four-way) marginal is included to account for changing demographic composition in the labor force. The third marginal configuration allows for period changes in the way that the age-sex groups experience underemployment. Finally, the (RLT) marginals in the configuration denote the changing relationship of race group to underemployment with time. The model in question yields $L^2 = 684$ on 495 df, with $L^2/df = 1.38$. Ordinarily, it is desirable to obtain a model with L^2/df approximately equal to 1 (Haberman, 1978, Chapter 1), but in the present situation our model fit is nevertheless acceptable. Adding more complicated interactions does not improve the fit very much in comparison to the sacrifice in degrees of freedom, and the large sample size (approximately 815,000) indicates that the model is performing remarkably well.

Parameter estimates for this model capture all of the important features of the data. In particular, parameter estimates that involve time serve to capture changing relationships over time; parameter estimates that involve both underemployment and time describe changing levels of underemployment over time. It is easiest for us to describe the changing relationship of race to LUF over time (because only the lambda parameters for the R-L-T interaction terms are relevant, see Appendix 3.1). Figure 3.5 presents these estimates for

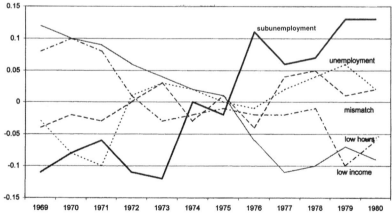

Figure 3.5. For blacks, growth in the proportion of unemployed, discouraged, and mismatched workers; declines in part-time and low-income employment (parameter estimates). Model fits marginals (ASRL) (ASRT) (ASLT) (RLT), $L^2 = 684$, 495 df, $L^2/df = 1.38$. Source: Clogg & Sullivan, 1983, p. 139, Table VIII.

blacks (estimates for nonblacks have the opposite sign). A negative value indicates that blacks are less likely, in a given year, to experience a particular type of underemployment than in other years.[5]

The interpretation of values in Fig. 3.5 is straightforward. Blacks were more likely to be discouraged workers with the passage of time. Except for 1976, blacks also were more likely to be unemployed later in the series. The trend in the mismatch parameter estimates indicates that blacks were increasingly likely to experience educational mismatch toward the end of the decade. However, blacks did noticeably better over time on hours and income measures of underemployment.

The trend in these estimates is fairly clear. Through the 1970s blacks were more likely to be discouraged workers and unemployed but less likely to be disadvantaged with regard to hours of work or income from work.

Discussion

Full employment has been an official government goal since 1947. Nevertheless, the level of employment, the indicator currently reported, bears no necessary relationship to social welfare because some workers are inadequately or marginally employed. The LUF can be applied to standard labor force surveys to produce much additional labor force information. As a social indicator, the LUF clarifies the relationship between employment and social welfare, fulfilling one objective of indicators of employment quality (Land, 1976, p. 22; Seashore & Taber, 1976, p. 122).

The LUF strategy is well suited for social indicators research in at least two fundamental respects. First, it is an objective social indicator, measuring the prevalence of the various types of underemployment identified over the past three decades. Second, it is also suited to the construction of normative social indicators, which gauge how well the labor market performs in distributing rewards to significant social groups. The LUF is easily adapted to include normative cutoff points, such as poverty thresholds, in its construction.

An example of the normative use of this indicator is evident in our analysis of the race–LUF relationship over time. Reduction in race-based differentials in underemployment is surely a goal of employment policy, and detailed analysis of the LUF × race × time interactions (Fig 3.5) shows how and where race differentials are moving toward more favorable circumstances for blacks. Trends in mismatch rates, which record a striking increase in the prevalence of mismatch over the 12 years studied here, are also of direct policy significance. Whatever the criticisms of the mismatch definition, there can be little question that the education–occupation relationship has changed rapidly, and that this potentially has significant consequences for worker expectations and satisfaction as well as possible consequences for productivity.

WHAT CAN COHORT ANALYSIS TELL US ABOUT RECENT
TRENDS IN LABOR FORCE PARTICIPATION?

Apart from trends in labor force activity by gender and age, most policy analysts are interested in whether there have been changes in these relationships over time. Specifically, most analysts are interested in separating *cohort effects* and *period effects* in labor force participation. Cohort effects are changes in labor force activity that occur because groups of people of similar age live through the same experience. This experience affects this group of people *and only them*—others at different ages are not affected in the same way (or not at all). Period effects are changes that occur because of unique historical experiences that happen to everyone who lived through them.

Some examples of cohort effects would be declining labor force participation rates among those who were 30–45 years old during the Depression of the 1930s. This group of prime working-age men was uniquely discouraged by the experience of the Depression and this effect was visible throughout their working lives.

An example of a period effect would be the higher participation rates of labor force entrants who lived during World War II. For this unique historical period, the labor force participation rates of all groups rose substantially.

The separation of age, period, and cohort effects have definite policy implications. The fundamental questions policy analysts often ask require that the analysis clearly separate these effects. For example, is the recent decline in labor force participation among black men indicative of an age effect (the concentration of black men in age groups that are least attached to the labor force), a cohort effect (the result of unique social and cultural influences affecting current black men as they were growing up), or a period effect (the result of distinctive historical opportunities unavailable now that were available in the past that affect all blacks)? (See Appendix 3.2 for further discussion.)

The objective of this section is to present a comprehensive view of recent trends in labor force participation by considering men, women, blacks, and nonblacks. The reference period is the 10-year interval from 1969 through 1979. The primary concern is with estimating the "intrinsic" cohort tendencies to participate in the labor force, and then examining the way that cohort differences condition the period change we observe.[6] Cohort effects can play a large role in labor force trends. We also conclude that the conventional discouraged worker hypothesis cannot provide a convincing explanation of the way that labor force participation fluctuates during times of recession or growth.

58 Chapter 3

A MODEL WITH AGE-PERIOD INTERACTION

The basic model (described in Appendix 3.6) allows for main effects of age, period, and cohort but it does not allow for interactions among these variables. The model thus makes assumptions that are worth elaborating more carefully, at least insofar as it is to be applied in the labor force participation context.

1. *The age effects, which indicate the age-graded character of labor force participation, or "life cycle" effects* (see Winsborough, 1978), *are assumed to be constant across periods and cohorts.* The basic model, therefore, makes no allowance for the possibility that period-specific shocks alter the participation behavior of different age groups in different ways. This assumption is not consistent with notions about the changing volume of discouraged workers (see Flaim, 1972) who are outside the labor force. Discouraged workers are principally composed of the young and the old; their numbers fluctuate considerably with economic growth and recession. Using the standard definition of the labor force implies that measured participation rates will fluctuate in proportion to the changing numbers of discouraged workers. Retirement patterns also change with time, implying that a special allowance should be made for age–period interaction for older workers. Decisions to retire might be influenced by general labor market conditions (for example) and this implies an interaction of age and period for older workers.

2. A second assumption is that the *period effects, which refer to the transitory labor market shocks as well as to other period-specific influences, are the same across age and cohort.* There is only one estimate that governs the entire influence of a given period's socioeconomic condition on participation and this parameter exerts its influence independently of age or cohort.

3. A third assumption is that the *cohort effects are assumed constant across age and period.* These cohort tendencies do not change as the cohort ages; they do not adjust to the peculiar period experiences that cohorts encounter as they age. A period–cohort interaction appears to describe what Durand (1948, pp. 123–124) had in mind when discussing changing participation patterns of women subsequent to the Second World War. His conjecture was that the abnormally high work rates of young adult women during the war continued across the life cycle, leading to increased rates of participation in later ages. For our relatively short time interval we can think of no peculiar period shocks as dramatic as the Second World War, and so we make no attempt to incorporate period–cohort interactions into a model. We can think of no reasons for believing there are age–cohort interactions either. However, the possibility of age–period interactions cannot be ignored because of historical fluctuations in the numbers of young and old discouraged workers and changing retirement patterns.

Though difficult to add to such models, an age–period interaction is required here to take proper account of the way that participation is responsive to labor market conditions. Conditions of "full employment" should increase measured participation at both ends of the age range, while recessions should decrease participation among the youngest and oldest workers. We also suspect that retirement trends are responsive to current economic conditions. Details regarding the construction of this model can be found in Appendix 3.6.

RESULTS

Indices of Fit

Figure 3.6 presents the index of dissimilarity (D) for gender and racial groups (details of the log-linear models estimated, df and L^2 are presented in Appendix Table 3.6). When the sample sizes are taken into account, it is clear that the model with main effects and age–period interactions performs rather well, although the participation experience of blacks is not accounted for as well as the experience of nonblacks (D = 0.66%, 1.17% for nonblacks, but D = 2.91%, 3.38% for blacks, for the age–period interaction model).[7] We believe this constitutes sufficient evidence to suggest that age, period and cohort each

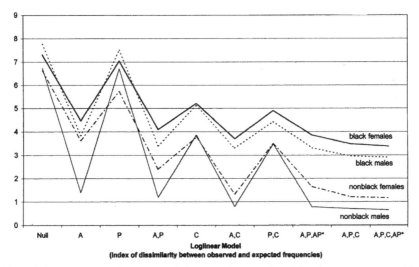

Figure 3.6. An age–period interaction helps to explain trends in labor force participation. The labor force participation of blacks is explained less weel than others.

exert effects on labor force participation, and, combined with the argument of the preceding section, the age × period interaction is required as well.

Estimated Cohort Effects

Figure 3.7 presents the estimated cohort effects for each of the gender–race groups (Appendix 3.7 discusses the identification restrictions used to produce these results). The quantities measure the effect of membership in a given cohort when contrasted with membership in the oldest cohort (using a logarithmic scale).

For both nonblack males and females, younger cohorts have a greater intrinsic tendency to participate than older cohorts, with an average intercohort change of +0.048 and +0.078 respectively. The cohort effects for blacks are not as reliable because of the smaller samples and the less satisfactory performance of the models. Nevertheless, these results show that younger cohorts of black males have less intrinsic tendency to participate and younger cohorts of black females have a greater intrinsic tendency to participate. The average intercohort change is –0.189 for black males and +0.233 for black females.[8] In view of the magnitudes of these effects, which are separate from age, period, and age–period effects, the conclusion must be that cohort differentiation exerts a strong influence on what is observed as across-time change.

How Cohort Effects "Translate" the Observed Period Change

Ryder (1964) referred to the "demographic translation" of cohort rates into period rates. In this section a simple method is used to show how the

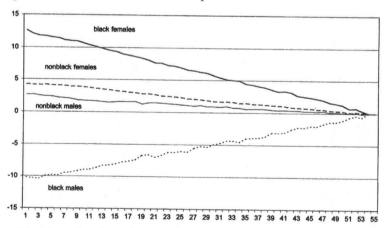

Figure 3.7. Cohort effects (logarithmic scale) from a model incorporating age–period interaction (labor force participation relative to the oldest cohort = 55). Source: Clogg, 1982a, *Demography*, p. 469, Table 3.

cohort effects "translate" the period effects, or equivalently, how the cohort effects confound the interpretation that might be given to the "crude" period rates actually observed. These rates depend on all of the effects in the model, except for the cohort effects, and so in this sense they indicate the "true" period-specific influences on participation. The observed period-specific participation rates (the "crude" rates) can be compared with the adjusted rates, thereby inferring the components of period change that are due to period effects and to cohort effects. This procedure is merely a means of assessing the effects of period and cohort in terms different from the odds and logits, and in terms consistent with "components" methods of demography (see Wunsch & Termote, 1978).

The successive differences of both the observed period rates and the adjusted rates are compared in Table 3.1. For nonblack males, the observed participation rate declined in 8 of the 10 years from 1969 to 1979. The average interperiod change was −0.2%. When the cohort effects are removed, the interperiod change is usually much more negative, averaging −1.7%. Thus, the cohort effects offset the effects of the period-specific influences; the combined influence of both kinds of effects operated to modestly reduce the participation of nonblack males. The average interperiod change observed minus the average "purged" interperiod change yields an estimate of the general influence of the cohort effects; this quantity is −0.2% − (−1.7%) = 1.5%. Thus, *if there were no period effects at all, nonblack males would increase their participation*

Table 3.1
Changes in Labor Force Participation Over Time Are a Function of Period and Cohort Effects, Not Just Period Effects Alone

Subgroups			Mean period change
Nonblack men			
Observed period change			−0.2
Period change purged of cohort effects			−1.7
Cohort effect	[−0.2 − (−1.7)]	=	1.5
Nonblack women			
Observed period change			1.1
Period change purged of cohort effects			−.6
Cohort effect	[1.1 − (−0.6)]	=	1.7
Black men			
Observed period change			−0.5
Period change purged of cohort effects			0.3
Cohort effect	[−0.5 − (0.3)]	=	−0.8
Black women			
Observed period change			0.5
Period change purged of cohort effects			−0.3
Cohort effect	[0.5 − (−0.3)]	=	0.8

Source: Clogg, 1982a, *Demography,* p. 471, Table 4.

on the average by 1.5% per year. Alternatively, if there were no cohort effects, they would decrease their participation on the average by –1.7% per year.

For nonblack females, the observed interperiod change is positive everywhere (averaging 1.1%). However, when the cohort effects are removed the interperiod change averages –0.6%. The average period-specific influences (–0.6% per year) are thus offset by cohort influences which average 1.1% – (–0.6%) = 1.7% per year.

Table 3.1 also presents results for black males and females. Over the entire interval of time covered by our data, the observed participation rate of black males declined by an average of 0.5% per year, while the observed participation rate of black females rose on average by the same amount. However, the data purged of the cohort effects are quite different; the actual influence of period-specific effects averages +0.3% for black males and –0.3% for black females. Thus the observed period change is conditioned to a great extent by cohort effects for these groups as well.

The results of Table 3.1 demonstrate the importance of a sociodemographic perspective in the analysis of recent trends in labor force participation. For each group, but especially for nonblack groups, the effects of cohort differentiation with respect to "intrinsic" labor force attachment must be reckoned with in order to give a proper accounting of across-time change. The usual economic view that attributes the bulk of across-time change in participation to labor market factors of supply and demand—period effects broadly conceived—is simply not supported by our analysis. Cohort effects, which by definition are not attributed to aging or period effects, must be considered to understand the very nature of recent trends. According to these results, an analysis of recent trends that incorporates only period-specific causal agents is bound to be misleading.

Period Shocks on Participation for Young and Old Age Groups

Our results so far assume that the effects of social and economic conditions at specific times on labor force participation vary by age. Such a model appeared to be consistent with notions about discouraged workers (see Flaim, 1972; Levitan & Taggart, 1974; Sullivan, 1978), as these marginal workers are principally composed of young and old "nonparticipants" (those outside the labor force). It also appeared to be consistent with trends in the age at retirement, if we accept withdrawal from the labor force in ages 50–64 as indicative of retirement.

Figure 3.8 presents the effects of "period shocks" on labor force participation for those usually considered most vulnerable to changing current labor market conditions. In this plot, the "young" graph describes the effect of a specific period on the labor force participation of a 20-year-old relative to a

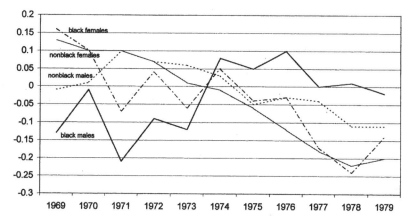

Figure 3.8. Period shocks operate to significantly reduce the labor force participation of older workers. Entries are the log odds of participation of a 64-year-old versus a 50-year-old. Source: Clogg, 1982a, *Demography*, p. 473, Table 6.

34-year-old. The "old" graph describes the effect of a specific period on the labor force participation of a 64 year-old relative to a 50-year-old. A positive number indicates that the period shock was favorable to the participation of the youngest and oldest cohorts, a negative number indicates the period shock was unfavorable to the participation of the youngest and oldest cohorts.[9]

For nonblack males, nonblack females, and black females, after 1974 or 1975 *the period shocks operate to reduce the participation of older workers.* That is, from about the middle of the decade onward the socioeconomic conditions of each period encouraged earlier withdrawal from the labor force for older workers for these graphs. Taking the quantities in Fig. 3.8 as a whole, it is clear that

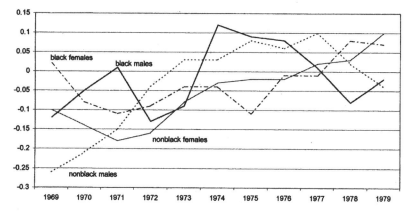

Figure 3.8. Continued. Period shocks on labor force participation are larger for young black men and men generally.

the period shocks on the older age group alone account for variability in age-period-specific participation rates from about +4% to −4%.

Note that the period shocks on the participation of older black males do not exhibit the same patterns as those for the other groups. Indeed, it is difficult to formulate any general statement about the trend in the period shocks for older black males. The lack of period-specific shocks might be the counterpart of large cohort effects.

Next, consider the quantities describing the period shocks on the participation of younger people. Our initial hypothesis was that these should fluctuate with economic growth and recession. For the time series considered here, 1969 best represents a "full employment" economy (unemployment was around 3.5% for the year), and 1974–1975 best represents recession experience (unemployment was around 8–9% for this interval).

Figure 3.8 suggests that our initial expectation is false: Employment opportunity of the several periods simply does not correlate with the estimated period shocks. For example, 1969—a full employment year—was actually unfavorable to the participation of 20-year-olds in comparison with 34-year-olds for three of the groups, and it was only negligibly favorable to the participation of younger people for the remaining group (black females). During the 1974–1975 recession, the period shocks were actually favorable to the participation of younger males, but they were unfavorable to the participation of younger females generally.

At least three different explanations for these findings can be offered. First, while it is true that younger people are more susceptible to being discouraged workers, probably the majority of discouraged workers are actually located in the 16–19 age category, a range of ages that was deliberately excluded here (see Clogg, 1979, Chapter 2). Second, the data pertain only to civilians, and the inferences drawn here would only be appropriate if excluded military personnel were representative of civilians. In the initial years of our series, there was probably a net movement of military personnel into the civilian work force, and this would confound the inference in complicated ways.[10]

Third, the "additional worker" hypothesis (see Bancroft, 1958; Long, 1958, p. 181) might actually be a better description of how young people as a group react to economic growth and recession. This hypothesis states that a certain segment of the population exists as a "ready reserve," forced into participating when recessions erode the economic well-being of those on whom they depend. This group is easily separated from the work force when conditions of full employment return. When a recession occurs, there are surely those who withdraw from the labor force because they have been discouraged by job seeking, but there are also those who are forced to enter the work force to "make ends meet" in their households.[11] In considering the recession of 1974–1975, *it appears that there were more additional workers than discouraged workers*

*among males, and that there were more discouraged workers than additional work-
ers among females.* Nearly the largest effects occurred during the relatively pros-
perous years 1969–1970. By virtue of the negative values of these effects for
the first three groups (–0.26, –0.10, –0.12, for 1969), the inference would be
that the conditions of full employment actually led to less favorable participation changes for the young relative to middle-aged. These results beg for a
more comprehensive explanation, as the discouraged worker concept simply
does not explain them.

DISCUSSION

This section has applied the age–period–cohort accounting framework
to analyze recent trends in labor force participation. The substantive conclusions were that (1) cohort differences are not trivial, (2) cohort effects play a
major role in producing the time trends observed, and (3) a convincing causal
analysis of recent trends cannot possibly be conducted with reference to only
period-specific causal variables such as measures of current labor supply and
demand. In these concluding remarks, certain criticisms of the methods used
will be addressed, some implications of our findings will be drawn out, and
suggestions will be made concerning how strictly demographic concepts might
be used to explain the results.

Our principal finding is that cohort effects and period effects (including
age–period interactions) had just the opposite influences on time trends for
each subgroup. For nonblack males period effects operated to decrease participation by an average of 1.7% per year, while cohort effects operated to
largely offset the period effects, producing an observed time trend that showed
a slight reduction in participation over the period. For nonblack females, period effects also operated to decrease participation (by an average of 0.6% per
year), but cohort effects more than compensated for the period effects, producing the increase in participation that was actually observed. For nonblack
males the period effects were actually much more negative than the observed
time trend would indicate, while for nonblack females the period effects were
actually the opposite of that indicated by the observed time trend. At the very
least, these results can be used to justify the use of a sociodemographic perspective (incorporating ideas of intrinsic cohort differentiation) in the analysis of recent trends. A narrow economic view, geared toward explaining trends
in terms of supply–demand conditions in the labor market, is impossible to
justify if cohort factors are considered.

Our results also suggest that, with the exception of black males, older
workers were withdrawing from the labor force earlier by 1975. Whether or
not these period shocks were largely caused by labor market conditions or by

social factors is a question that our data and methods cannot address. We showed that these period shocks had nontrivial effects on participation, which is consistent with findings reported by O. D. Duncan (1979).

We found unanticipated results when estimating the period shocks on the participation of the young age group (20–34). The discouraged worker hypothesis was not consistent with the estimated pattern of effects, and we suggested that the additional worker hypothesis might better explain the findings for some groups. Clearly, more research needs to be done on the way that labor market conditions affect the labor force activities of young people.

SHOULD UNDEREMPLOYMENT RATES BE ADJUSTED?

We next turn to the question of whether underemployment rates should be adjusted to account for the composition of the labor force. Most people are interested in whether the labor market is providing adequate employment opportunities. Most people are also interested in how potential employers adjust to the loosening and tightening of labor markets. But comparisons over time are not as easy as they appear to be. An economy with high proportions of young and old people (those least likely to be connected to the labor market) will have a different pattern of underemployment than an economy dominated by 30- to 45-year-olds. Only part of the difference between these economies will be due to differences in the opportunities actually available. The economy with a large subpopulation of those weakly attached to the labor market may have an abundance of current and potential opportunities. What is lacking are people in the optimal life course stage to take advantage of them. The economy dominated by prime working-age people may lack suitable opportunities for those ready to take advantage of them. These complexities might be glossed over if unemployment measures were collected and immediately compared across time periods. Such arguments have led many researchers to argue that unemployment rates need to be adjusted to take into account differences in the composition of the labor force (see Gordon, 1976; Taueber, 1976; Smith, 1977; Easterlin, 1978; Clark & Menefee, 1980; Wernick & McIntire, 1980).

The common theme in these arguments is that *unemployment rates typically vary across demographic groups.* For example, labor force members aged 16–24 have unemployment rates that are two to four times greater than the rates observed for those aged 35–40. Second, *the demographic composition of the labor force has changed substantially since the Second World War* (see Durand, 1948; Bancroft, 1958; Long, 1958; Sweet, 1973; Clogg, 1979, 1982a). For example, the age distribution of the labor force from 1950 onward changed in part because of the Depression lull in fertility and the post-War baby boom.

Third, *unemployment rates should be adjusted for compositional changes to esti-mate the "true" temporal change in unemployment that arises solely from changes in a set of age–gender-specific unemployment rates.* Using such logic, Antos, Mellow, and Triplett (1979) suggest that the 1977 crude unemployment rate (about 8%) was 1 to 1.4% higher than when the 1957 labor force composition was used as a standard for adjustment. Compositional changes between 1957 and 1977 tended to increase the crude unemployment rate in 1977 by over 1%.

If we want to compare recent unemployment since 1970 to unemployment 20 or more years earlier, the available research indicates that composition adjustments must be made. However, this prior work does not demonstrate that composition adjustments are required when assessing:

- Short-term change in unemployment (over 2- to 5-year intervals)
- Medium-term change in unemployment (over a 10- to 15-year interval)
- Recent trends in unemployment (from the late 1960s into the 1980s)
- Change in underemployment, as distinguished from unemployment

Previous research leads us to expect that composition adjustments are required at least for assessing medium-term and recent change in unemployment. Because underemployment is closely related to unemployment in terms of its distribution across demographic groups, we could argue that composition adjustments are required for underemployment rates as well.

However, we show in this section that *it is not necessary to carry out composition adjustments in either crude rates of unemployment or crude rates of underemployment to analyze labor force behavior over the recent past.* Changes in the demographic composition of the labor force have had little to do with the secular increase in unemployment and underemployment from the late 1960s to the 1980s.

Compositional Change from 1970 to 1980: An Overview

Figure 3.9 presents the age distribution of the civilian labor force in 1970 and in 1980 taken from the March Current Population Surveys (CPS). We see a substantial proportionate increase in the labor force aged 20–34 over the decade, which was offset by a decrease in the proportion of the labor force aged 40–64. The magnitude of the net change is described by the index of dissimilarity, $D = 9.7\%$. A 10% net shift in the age distribution over a decade is substantial, and such a shift indicates that an adjustment for changing age composition should be considered before interpreting trends in unemployment and underemployment over the interval.

Not all changes in the age distribution were operating to increase unemployment and underemployment in 1980 relative to 1970. Unemployment rates, for example, are highest for the 16–17 age group, but there was virtually no

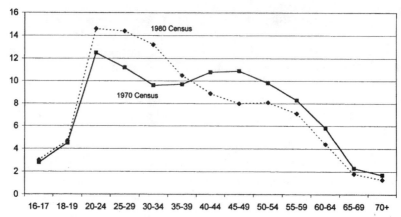

Figure 3.9. Labor force participation rose for 20- to 30-year-olds, but declined for 40- to 70-year-olds from 1970 to 1980. Entries are percentages. Source: Clogg & Shockey, 1985, *Demography*, p. 397, Table 1.

change in the proportional representation of this age group over the decade. Unemployment rates rise after age 55 but there was a net *decline* in the share of the labor force claimed by this group. The major change in the age distribution that favored an increase in unemployment was the proportionate increase in 25 to 34 year-olds. In sum, at least some of the changes in composition that tend to increase unemployment are compensated for by changes that tend to decrease it.

Figure 3.10 gives the change in the percentage of each labor force age group composed of women, based on Census estimates. Overall, female labor

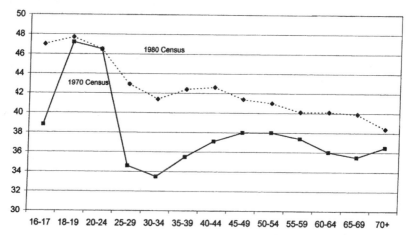

Figure 3.10. Women's labor force participation by age. The rise of women's labor force participation is not uniform across age groups. Entries are percentages. Source: Clogg & Shockey, 1985, *Demography*, p. 399, Table 2.

force participation increased from 41.3% to 49.8% between 1970 and 1980, and women's proportion of the total labor force increased from 38.1% to 42.6%. But this increase was by no means uniform across age groups. The percentage of women increased for all age groups except 20- to 24-year-olds. The increase was greatest for women aged 25–44.

The net effect of such changes in the age-by-gender composition of the labor force on unemployment and underemployment rates is difficult to predict. Women as a group tend to have unemployment rates that average about 1% more than unemployment rates for men, but unemployment varies by age. The increase in the percentage of women for age group 16–17 would favor increasing unemployment, but the lack of change in this percentage for ages 18–24 where unemployment risk is great would operate to keep unemployment levels stable. The increases observed for ages 25–44 partly favor an increase in unemployment (especially for ages 25–29) but also partly favor an overall maintenance or even a decline in unemployment (for ages 30–44 where unemployment risk is low). Increases in the percentage of women aged 50 and over should increase unemployment because labor force members in this age range face a higher risk of unemployment.

On balance, the changes that occurred in the percent female on an age-by-age basis probably operated to produce a net increase in unemployment and underemployment, although the magnitude of the effect is something that cannot be predicted with any certainty.

Compositional Change as a Component of Overall Temporal Change

To say that there is compositional change is to say that there is an association between a compositional variable and the period in which it is observed. Conversely, a lack of composition change is denoted by statistical independence in a composition-by-period contingency table. In this section we assess the relative magnitude of the association between composition and period in contingency table terms. This gives us a means of evaluating the statistical importance of compositional change in the labor force.

Note first that the data to be adjusted are in the form of a contingency table. This table cross-classifies composition (age–sex categories: 52 levels), period (6 levels for EVEN YEARS and ODD YEARS, 12 levels for ALL YEARS), and the LUF variable (underemployment categories: 6 levels). The contingency tables are of dimension 52 by 6 by 6 (1872 cells) or 52 by 12 by 6 (3744 cells). Log-linear models that focus on the role of temporal change in the labor force are formulated for the tables (see Goodman, 1978; Haberman, 1978; Fienberg, 1980). Precedents for the approach taken here can be found in Clogg (1979, Chapter 3) and Clogg and Sullivan (1983).

Table 3.2 presents likelihood-ratio chi-squared values (L^2) for some log-

Table 3.2
Age Composition Exerts a Greater Effect
on Underemployment than the Gender Composition
of the Workforce

Model	Margin Fit[a]	L^2	df
H_0	(LAS)(P)	21,305	3421
H_1	(LAS)(LP)	12,454	3366
H_2	(LAS)(LP)(ASP)	5280	2805
H_3	(LAS)(ASP)	13,580	2806
H_4	(LAS)(LP)(AP)	7314	3091
H_5	(LAS)(LP)(SP)	11,191	3355

[a]L = LUF category; A=Age; S=Sex; P=Period
Source: Clogg & Shockey, 1985, *Demography*, p. 401, Table 3.

linear models applied to the three contingency tables formed from the CPS data. Model H_0 fits the marginals [(LAS), (P)], where L refers to the LUF variable, P to period, A to age, and S to sex. This model posits independence between period and the joint variable L-A-S. Including the LAS marginal configuration allows for all possible interactions among LUF, age, and sex. The parameters included by fitting this marginal take account of the fact that underemployment levels depend on age and sex. The chi-squared value for H_0 measures the total period effect on the labor force, or the gross amount of labor force change, because (1) the model excludes the interactions between period and any other characteristic, and (2) any more comprehensive model must introduce interactions between period and one or more of the other variables. The models that follow are used to partition the chi-squared value for model H_0.

Model H_1 fits the marginals [(LAS), (LP)] and differs from H_0 solely by allowing a period effect on the LUF variable. This model posits period change in underemployment but no other kind of period change. This model does *not* allow for changing demographic composition (interaction of P with AS) *nor* for changing underemployment that depends on age or sex (interaction of P with LAS). Comparing L^2 values for H_0 and H_1 isolates the chi-squared component "due to" period change in underemployment. Model H_2 fits the marginals [(LAS), (LP) (ASP)]. This model incorporates change in age–sex composition, as it allows for interaction of P with A, S, and AS. By comparing L^2 values for H_1 and H_2, we are isolating the component "due to" changing demographic composition.

The partitioning of chi-squared values suggested by the sequence of models H_0 to H_1 to H_2 is not unique, implying that the chi-squared components associated with either period change in LUF or period change in composition are not unique. An alternative route is to consider model H_3, which fits the marginals [(LAS), (ASP)]. This model allows for changing composition (A-S-P interaction) but does not allow for period change in underemployment [because H_3 does not fit the (LP) marginal]). When H_3 is used the proper sequence of models is H_0 to H_3 to H_2. The partitioning suggested by this sequence first "takes out" the component due to composition change, and after this the component due to period changes in underemployment. This partitioning strategy will tend to give more weight to compositional change, less weight to underemployment change. Models H_4 and H_5 in Table 3.2, which are nested between H_1 and H_2 (although not themselves nested), can be used to consider whether age compositional changes or sex compositional changes are more important in the overall age–sex composition change.

The relevant chi-squared components based on the partitioning sequence H_0 to H_1 to H_2 appear in Table 3.3.

Column 2 gives the percentage of the total period effect associated with changing demographic composition. There is a modest difference between results obtained from ODD YEARS versus EVEN YEARS (30.4% versus 37.6%); the corresponding quantity calculated from ALL YEARS is 33.7%. The proportion of the total period effect associated with period change in underemployment (line 1) varies even more substantially between ODD YEARS and EVEN YEARS (48.1% versus 35.8%). A possible explanation for this difference is that the odd years in the series happen to exhibit the most temporal variability in underemployment, and this is reflected in the relevant chi-square components. For example, 1969 had the lowest unemployment (3.4%) and 1975 the highest (9.0%). Because the components in lines 1 and 2 are obtained from a nested sequence which produces the additivity of chi-squared values, this might explain the ODD–EVEN difference in line 2 as well. But because

Table 3.3
Decomposition of Total Period Effect Based on Chi-Square Values

Type of Interaction	Chi-squared values used[a]	Even years		Odd years		All years	
		L^2	%	L^2	%	L^2	%
1. Period–LUF	$L^2(H_0) - L^2(H_1)$	3446	35.8	5,307	48.1	8,851	41.5
2. Period–Age–Sex Given 1.	$L^2(H_1) - L^2(H_2)$	3625	37.6	3,351	30.4	7,174	33.7
3. Residual[b]	$L^2(H_2)$	2559	26.6	2,374	21.5	5,280	24.8
4. Total	$L^2(H_0)$	9630	100.0	11,032	100.0	21,305	100.0

[a]Chi-squared values.
[b]The residual measures the three-factor interaction among LUF, period and age–sex composition.
Source: Clogg and Shockey, 1985, pg. 402.

the inferences seem to depend on whether ODD YEARS or EVEN YEARS are considered, we shall continue to analyze each file separately.

Somewhat different results are obtained if the partitioning sequence H_0 → H_3 → H_2 is used. Even though there are modest differences among the results for different tables and for different partitioning sequences, *the important fact is that compositional change by itself accounts for about one-third of the overall period effect on the labor force.*

Models H_4 and H_5 (see Table 3.2) can be used to examine the relative contributions of the age and sex compositional changes. The logic for partitioning is the same as above. The following results are obtained when considering the ALL YEARS file. If H_4 is used, we find that 71.6% of the compositional change is attributed to changing age composition, 28.4% to changing sex and/or age–sex composition "net" of changing age distribution. If H_5 is used we find that 17.6% of the compositional change is attributed to changing sex composition, 82.4% to changing age and/or age–sex composition "net" of changing sex composition. Changing age composition has been more important than changing sex composition according to these indices.

The results of the log-linear analysis (further detailed in Appendix 3.8) indicate that *compositional change has been substantial when gauged in relation to overall temporal change in the labor force.* This information cannot be used, however, to infer the magnitude or even the direction of the effect of compositional change on crude rates. Existence of composition–period interaction has been documented, but this is only a necessary condition for compositional effects on rates. As we shall argue below, much of the compositional change that favored increases in underemployment was offset by compositional change that favored decreases in underemployment.

Composition-Adjusted Rates

Table 3.4 presents crude and adjusted rates of underemployment for the earliest and latest year in each of the "all years" file.

On balance our results indicate that there is at most a negligible compositional effect on trends in underemployment over the 1969–1980 interval, which is in sharp contrast to what we anticipated. There is simply no appreciable compositional effect at all when considering discouraged workers. The 1969–1980 change in crude rates (0.00%) is essentially identical to the 1969–1980 change in the adjusted rates (0.05%).

Next consider the compositional effect on unemployment rates. The difference in crude rates (1980 rate minus 1969 rate) is 3.12%. The corresponding difference in adjusted rates is 2.87%. The difference between these two quantities is only 0.25%, which is our estimate of the compositional effect on unemployment rates over this interval. Note that the effect is positive as an-

Table 3.4
Crude and Adjusted Underemployment Rates

Year/ rate	Labor utilization framework category				
	Subunemployed	Unemployed	Low hours	Low income	Mismatch
Crude rates					
1969	1.18	3.40	2.39	8.22	7.80
1980	1.18	6.52	3.43	7.29	14.18
Difference (1980–1969)	0.00	3.12	1.04	−0.93	6.38
Rates adjusted for Composition*Year					
1969	1.19	3.56	2.39	8.12	7.83
1980	1.14	6.43	3.41	7.39	13.66
Difference	−0.05	2.87	1.02	−0.73	5.83
Rates adjusted for Composition*Year and Composition*LUF					
1969	1.23	3.43	2.42	8.27	8.25
1980	0.97	6.20	3.28	7.42	13.41
Difference	−0.26	2.77	0.86	−0.85	5.16

Source: Clogg & Shockey, 1985, p. 404, Table 5.

ticipated: The compositional changes that occurred tended to *increase* the prevalence of unemployment. But this effect is very small, certainly much smaller than would be expected from research results cited in the introduction.

Stated in relative terms, *the compositional effect is only 8% of the difference in crude rates between 1969 and 1980*. It can be verified that the composition effects on low-hours underemployment (H) and on low-income underemployment (I) are also negligible. The mismatch form of underemployment is a different case. The difference in crude rates was 6.38%, and we see that 0.55% of this can be attributed to compositional change. However, in relative terms this represents only 9% of the increase in mismatch over the interval.

Even though there has been much change in the demographic composition of the labor force between 1969 and 1980, adjusted rates that control for this change fail to show a substantial compositional effect on the crude rates or trends in them. The crude rates essentially speak for themselves whether the focus is on short-term or medium-term change in recent labor force experience.

It is conceivable that a more substantial compositional effect might be found if underemployment categories were suitably combined. Clogg (1979, Chaps. 2, 5) suggested condensing LUF categories of subunemployment, unemployment, low hours, and low income from employment into one category called *economic underemployment*, while retaining the mismatch and adequately employed categories. For economic underemployment as a combined category, the difference in crude rates (1980 rate minus 1969 rate) is 3.23%. The corresponding difference in adjusted rates is 3.12% so the effects of changing age and gender composition account for a scant 0.11% of the difference in rates. Thus, compositional effects are still trivial when underemployment categories recognized in the LUF scheme are condensed.

We next consider adjusting rates against "standard" compositions obtained from a stable population model. Stable age distributions for each sex were calculated from complete (single year) life tables for 1975 (National Center for Health Statistics, 1979, Table 5-2; see Keyfitz, 1968. Chapter 11) for the relevant numerical methods). Intrinsic growth rates (r) of +0.01, 0.00, and – 0.01 were used to create the age distributions, with the intermediate value corresponding to the stationary population. The age-specific labor force participation rates for each sex observed in the 1969 CPS were used to create the age distribution of the labor force.[12]

It seems clear that applying the 1969 labor force participation and underemployment rates to these stable age distributions increases the likelihood of finding compositional effects. The reason for this is that 1969 had the most favorable schedule of rates (labor force participation, unemployment, and underemployment) of any period in the series.

Table 3.5 summarizes the results for both compositional change and changes in rates. We assess differences by comparing the 1980 rate (crude or adjusted) with the crude rate of the standard. First consider results based on the stable population with positive growth ($r = +0.01$). The difference in crude unemployment rates (line 1.a) is 3.24% while the difference in adjusted rates (line 1.b) is 2.71%; the compositional effect is 0.53 percent in absolute terms, which represents 16% of the crude difference. In terms of variance (s^2 lines 2.a and 2.b) about 18% of the variance in crude rates can be attributed to a compositional effect. Similar compositional effects are observed for sub-unemployment and low-hours underemployment; low-income and mismatch underemployment types fail to exhibit compositional effects here.

When a stationary age distribution is used ($r = 0.00$). more pronounced compositional effects are obtained. For unemployment rates (lines 3.a and 3.b), the compositional effect is 1.0%. which is consistent with the compositional effect produced in previous research using standards obtained from the 1950s. About 29% of the difference in crude rates can be attributed to the effect of compositional differences in this case.

Compositional effects are even more pronounced when the negative growth ($r = –0.01$) stable age distribution is used as a standard. For example, the absolute compositional effect on unemployment rates is 1.4%. Probably this figure represents an extreme upper bound on the compositional effect, as

Table 3.5
Differences in Crude and Adjusted Rates (%) and Across-Time Variability in Rates, Using Stable Populations as Standards[a]

| | Labor utilization framework category | | | | | |
	Subunemployed	Unemployed	Low hours	Low income	Mismatch	Adequate
Stable population, $r = 0.01$ as base[b]						
1.a Differences in crude rates, 1980–standard[c]	.09	3.24	1.10	−0.62	6.10	−9.90
1.b Difference in adjusted rates	−0.04	2.71	0.97	−0.70	5.95	−8.89
2.a s^2, crude rates	0.069	2.85	0.318	0.254	5.13	14.68
2.b s^2, adjusted rates	0.056	2.35	0.273	0.290	5.13	12.12
Stable population, $r = 0.00$ as base						
3.a Differences in crude rates, 1980–standard[c]	0.12	3.47	1.08	−0.81	6.10	−9.97
3.b Difference in adjusted rates	−0.05	2.51	0.89	−0.81	5.81	−8.34
4.a s^2, crude rates	0.074	2.96	0.317	0.281	5.13	14.74
4.b s^2, adjusted rates	0.054	2.12	0.252	0.332	4.78	10.62
Stable population, $r = -0.01$ as base						
5.a Differences in crude rates, 1980–standard[c]	0.14	3.68	1.06	−1.09	6.15	−9.93
5.b Difference in adjusted rates	−0.06	2.31	0.81	−0.94	5.62	−7.73
6.a s^2, crude rates	0.071	3.06	0.314	0.332	5.15	14.70
6.b s^2, adjusted rates	0.055	1.93	0.233	0.392	4.38	9.12

[a]Adjusted for the composition–year interaction.
[b]See text for a discussion of the base populations used.
[c]difference between 1980 crude rate and crude rate of the standard.
Source: Clogg & Shockey, 1985, p. 410, Table 9.

a stable age distribution with $r = -0.01$ is very different from observed age distributions during the 1969–1980 interval.[13]

Our results raise important questions, especially because they tend to produce more substantial compositional effects than earlier results that were not based on external standards. First, should we adjust underemployment rates against an external standard, or should we rely on only the information about compositional change and rate change in the interval under study? The choice has an effect on the magnitude of the compositional effect that will be estimated. Second, if we use an external standard, *which one* should be selected? Inferences depend on the standard that is selected, as results in Table

3.5 show. Third, what time referent is more relevant for demographic research on the labor force or for public policy? Should we focus on long-term comparisons (covering a span of 20 years or more) or on short term comparisons drawn from recent experience?

SUMMARY AND DISCUSSION

Although there has been much change in the age–gender composition of the labor force since the late 1960s, the net effect of these changes on crude rates of unemployment and underemployment has been negligible. The mismatch type of underemployment is a partial exception to this but even here the relative effect of changing composition is modest. These conclusions contrast sharply with those given in or implied in prior research. The reason for the contradiction is that the earlier research concentrated on long-term comparisons or were based on the use of external standard compositions. Results presented here pertain to recent trends and short- to medium-term comparisons of crude and adjusted rates; they were obtained by adjusting trends in the 1969–1980 interval for compositional changes that occurred in exactly the same interval.

In assessing the trends in unemployment and underemployment from the late 1960s to the 1980s, we have found that the crude rates speak for themselves: There is no need to adjust these rates for changing demographic composition of the labor force. The secular rise in unemployment and most types of underemployment throughout this interval had little to do with compositional effects. Thus, the prevailing view that a large part of the "underemployment problem" since the early 1970s has been due to compositional change of baby boom cohorts entering the labor force and increasing female labor force participation is simply not consistent with the evidence.

We were astonished to find such small compositional effects, especially given the evidence in previous literature for such effects. In fact, we began this research thinking that compositional effects would be more substantial than had been estimated in previous work.[14] If various external standards that differ appreciably from recent labor force compositions are used, then it is possible to find more substantial compositional effects. But it must be emphasized that there is no inherent justification for using such external standards, especially if the focus is on recent trends.

We have said very little about *why* compositional effects are so negligible, why such substantial change in the demographic composition of the labor force tended to have such benign effects on unemployment and underemployment. This question needs to be answered by examining the detailed information on change in the age–sex composition. The results given here make it obvious that the various compositional changes that took place tended to compensate for one another, so much so that the net effect of all such changes on unemployment and underemployment was virtually nil.

ACKNOWLEDGMENTS

Research presented in this chapter was supported in part by Grant No. SES-7823759 and SES-8303838 from the Division of Social and Economic Sciences of the National Science Foundation. Prior published versions of these analyses appear as "Labor Force Composition and Underemployment Trends, 1969–1980," *Social Indicators Research* 198312:117–152 (co-authored with Teresa A. Sullivan); "Cohort Analysis of Recent Trends in Labor Force Participation" *Demography* 1982 19:459–479; and "The Effect of Changing Demographic Composition on Recent Trends in Underemployment," *Demography* 1985 22:395–414 (with James W. Shockey). Tables from these published works appear with permission of the publishers.

NOTES

1. Subjective measures of mismatch are much higher. In a survey of workers currently employed at least 20 hours a week, 21.6% strongly disagreed with the statement, "My job lets me use my skills and abilities" (Quinn & Staines, 1979, p. 194). Among workers reporting work-related problems, 35.6% indicated that "underutilization of skills" was a problem; 32.2% reported that "overeducation of the worker" was a problem (Quinn & Staines, 1979, p. 292).

2. Figure 3.1 does *not* document changes in "poverty status" of households or people because the definition of the income component is specifically related to earnings acquired in the labor market, not total household income (which includes nonwork income).

3. The change in subunemployment ratios between men and women is driven by changes in discouraged workers among men and women. The male rate jumps from 0.7 to 1.2% from 1969 to 1972, then declines gradually to a low of 0.6% in 1980. The women's rate jumps from 1.9% in 1969 to 2.4% in 1972 and is never below 2% after that.

4. We should note that this is not a cohort phenomenon, because those aged 20–34 in 1969 were only 31–45 in 1980. Nor can it be due to increased educational attainment: Those aged 50–64 in 1980 completed their schooling prior to the periods involved in this study

5. In this analysis we assume no age–period interactions for the prime working ages 35–49. (We did consider another specification where i′ = 29 and i″ = 54, and we found results that differed only slightly from this specification.) The choice of these ages is warranted, in our view, because the 35–49 age range is precisely that where unemployment is the lowest, where job security is the greatest, and where discouraged workers are virtually nonexistent. Workers in these ages are typically those with much experience, and their job skills are not yet outmoded. On the other hand, labor force participation is still in the process of being established for many people under age 35, and it is in the process of being terminated for many people over age 49. The assumption is then that the transitory period shocks on participation will exert their influences in different ways for each of the three age groups (20–34, 35–49, 50–64).

6. The term "intrinsic cohort tendency" is used repeatedly in this paper. It refers to the tendency to participate in the labor force (expressed in terms of odds or log-odds) that can be described solely as a cohort attribute. It cannot be attributed to age or period influences of the age groups or periods considered. (Intrinsic cohort tendencies could, however, be partly due to the age, period, or age–period influences of age groups or periods prior to the age groups or periods under study. Thus, cohort tendencies inferred in an analysis of 1969–1979 data might actually be due to lagged period effects, i.e., influences of period-specific causal variables operating prior to 1969).

7. Nonblack nonwhites constitute only about 10% of the nonwhite figure. No attempt was made to isolate Hispanics for separate treatment.

8. The magnitudes of these numbers seem suspect, but we do not see how different plausible values of (Δ) would change the size of these numbers.

9. These assumptions imply, e.g., that a 20-year-old is more susceptible to a period shock than a 34-year-old, that a 64-year-old is more susceptible than a 50-year-old. Such a specification is similar to Clogg's (1979, Chapter 9) model of "constant age elasticity," and it merely states that participation is more transitory at the ends of the age distribution (Clogg, 1982b; Goodman, 1979).

10. This cannot suffice to completely explain the findings in part because it does not explain the unexpected findings for women, and because it does not explain the unexpected findings after the mid-1970s when movements between military and civilian life were less consequential.

11. Appendix 3.8 gives a method of rate adjustment that controls for both the composition–period interaction and the three-factor interaction among composition, period, and the underemployment variable. The existence of three-factor interaction can complicate rate adjustment procedures (see Kitagawa, 1966; Bishop et al., 1975, p.133). The results just discussed can thus be questioned because they do not take account of the three-factor interaction in the data (i.e., the interaction of AG, P, and L). We next discuss adjustments for the three-factor interaction with the objective of validating the above inferences.

12. Because the rate adjustment methods employed here require a complete cross-classification of the composition categories by the categories of the dependent variable (LUF), the 1969 LUF distribution was used to create the hypothetical age-by-LUF "standard" that is the reference group for the adjustments.

13. These comparisons were based on taking differences between the 1980 crude (or adjusted) rates and the crude rate for the contrived standard. When differences are calculated between the 1980 crude (or adjusted) rates and the 1969 crude (or adjusted) rates, still using the contrived stationary age distribution as a standard from which compositional effects are to be judged, more modest compositional effects are obtained. The absolute effects (difference in crude rates minus difference in adjusted rates) are –0.03% (subunemployment), 0.63% (Unemployment), 0.17% (low hours), –0.09% (low income), and 0.57% (mismatch). Thus, the inferences obtained in the previous section are about the same for the subunemployment, income, and mismatch types of underemployment as when the trend is adjusted against a stationary population standard. The compositional effects for unemployment and part-time unemployment are somewhat higher when the stationary population standard is employed. For example, results just given estimate the compositional effect on unemployment rates at 0.63%, as compared with 0.25% obtained in the previous section. It must be borne in mind that the standard used here almost guarantees that a more substantial compositional effect will be estimated.

14. Because we were so surprised by the results, many other adjustment strategies were used in addition to the ones reported here. None of them produced estimated compositional effects that differed substantially from those discussed here. We even thought that the rate adjustment technique (Clogg, 1978) might be at fault, and to check on this we applied more

conventional standardization methods (Kitagawa. 1964). There was no appreciable difference between results obtained from the more conventional methods and results obtained from our own. For example, when using the average age–sex composition of the 1969–1980 interval as a standard to which the rates of each period were adjusted, the compositional effect on unemployment rates was 0.34%, which is consistent with the estimates obtained here (0.25% to 0.35%).

APPENDIX 3.1. ESTIMATING LOG-LINEAR MODELS

The log-linear models estimated that form the basis for the conclusions presented on pp. 54 and 55 are presented in Appendix Tables 3.1 and 3.2 below.

Appendix Table 3.1
Hierarchical Log-linear Models Applied to the Age
(A) × Sex (S) × Race (R) × LUF (L) × Period (P) Contingency Table[a]

Model	Margins fit	df	L^2
H_0	(AGRL)(P)	1309	16,259
H_1	(AGRL)(PL)	1254	7,646
H_2	(AGRL)(TU)(ASCT)	1045	2,863
H_3	(ASCU)(ASCT)	1100	11,180

[a]The total sample size (adjusted for weighting) is approximately 815,000.
Source: Clogg & Sullivan, 1983, p. 136, Table V.

APPENDIX 3.2. DECOMPOSITIONS OF L^2 IN MODELS OF THE RELATIONSHIP BETWEEN AGE (A), GENDER (G), RACE (R), LUF (L), AND PERIOD (P). (SEE APPENDIX TABLE 3.1)

Appendix Table 3.2

	Model	Method A		Component
		L^2	L^2/L^2 (H_0)	
1.	H_0	16,259	1.000	Total period effect
2.	H_1	7,646	0.470	Period–group and period–group–underemployment
3.	$H_1 - H_2$	4,783	0.294	Period–group
4.	H_2	2,863	0.176	Period–group–underemployment
5.	$H_0 - H_1$	8,613	0.530	Period–underemployment

Source: Clogg & Sullivan, 1983, p. 136, Table VI.

The purpose of this analysis is to isolate the relationship between sex (S), age (A), race (R), period (P), and the LUF components (L). Model H_0 assumes that the relationship between the demographic variables (A, R, and G) and underemployment (L) is the same across time (i.e. independent of period). This model gives an L^2 that we view as the total effect of time or period in our analysis.

The second model (H_2) allows for a main effect of period on underemployment (PL). This addition improves the overall fit but it does not reduce the value of L^2 to zero, which suggests that the entire period effect is not captured by the relationship between period and underemployment.

Other types of period effects are possible, and the remaining models examine these effects. Specifically, demographic composition could vary by period (AGRP) or the effects of demographic composition on underemployment could vary by period. Model H_2 adds (GRP) to model 1, and the improved model fit suggests that the interactions of demographics with change over time are large. But as this model still does not fit the data, there must be a higher-order interaction that is important. In short, there are marked differences in the way that the age–gender–race groups change in their risks of underemployment over time; business cycle effects (as represented by period) change the prospects of different groups.

Model H_3 allows for change in demographic composition over time (AGRP) but not changes in underemployment over time (no PL term). This provides another gauge of the importance of changing demographics on underemployment.

These models lend themselves to several different methods for isolating the effects of change over time in the relationships between demographic characteristics and underemployment categories. The top two rows of Appendix Table 3.2 report L^2 values for the models in Appendix Table 3.1. The remaining rows isolate specific effects using different logics of decomposition. Each method attributed 52% of the period effect to the main effect of time period change on underemployment (PL, see row 6 under Method A and row 3 under Method B). The interaction of period with demographic groups (AGRP) accounts for around 30% of the total period effect (row 3, Method A and row 6, Method B). Finally, row 4 in each half of Appendix Table 3.2 isolates the interaction between age, gender, race and the LUF categories that reflect differences in the ways that demographic groups experience underemployment through time. Around 17% of the overall period effect is explained this way.

The results in these models lead to our conclusions on p. 54;

1. Demographic change accounts for around 30% of the effect of changing time period on underemployment.
2. Another 18% of the period effect is explained by changes in underemployment rates that are tied to specific groups.

APPENDIX 3.3. RATIONALE FOR AND DEFINITIONS
OF LOW INCOME AND EDUCATIONAL MISMATCH
COMPONENTS OF THE LUF

A problem with the earlier application of LUF (in Clogg, 1979, see Appendix 1 of Chapter 2) was the difficulty in replicating the operational definitions. This appendix provides a rationale for procedures used to define the income and mismatch indicators used in this chapter. These procedures are easier to use and produce results that are similar to those using more complicated methods.

In both Sullivan (1978) and Clogg (1979), much work was necessary to obtain the low-income measure of underemployment. Clogg used information on *family* or *household* income for determining the income cutoffs used for primary earners in households, and complicated procedures were used to arrive at cutoffs for nuclear families contained within extended families or households. Different criteria were applied to men and women, to young and old, and to those in urban and rural areas, and complicated rules were used to provide the criteria for single or "unrelated" individuals. Although this complicated procedure corresponded to variations in the official poverty threshold, it was difficult for others to replicate (see the very detailed definitions in Clogg's Appendix A). Clogg also assumed that quite different income thresholds should be used to measure work adequacy of workers who differed only in terms of residence, gender, age, or relation to the primary earner in the household. This assumption makes sense for an indicator of economic welfare, but not for an indicator of adequate employment.

To simplify the choice of an income cutoff, we chose an income figure that produced a reasonable association between income level in the preceding year and current unemployment. (For a precedent, see Clogg, 1979, pp. 81–118.) Appendix Table 3.3 presents unemployment rates by income category for selected years and demographic groups.

The year 1969 was a year of full employment, with a March unemployment rate of 3.4%. The poverty threshold and previous year's work-related income were adjusted to a weekly wage, using number of weeks worked. The proportion of the labor force with incomes between the poverty threshold (PT) and [1.25 × PT] had 6.4% unemployment, an absolute difference of 3.1%. As the table indicates, the change from 1.00 × PT to 1.25 × PT produces a greater absolute change in unemployment risk than any other comparison of adjacent income categories. This result is true in both 1969 (a full employment year) and 1975 (a recession year with over 9% unemployment), regardless of race or gender, and for all age groups for 1975. (There are some anoma

Appendix Table 3.3
Unemployment Rates by Income Category, for Selected Years and Groups

Group	Income category (poverty cutoff X _____)				
	1.00	1.25	1.50	1.75	Total—all groups
1969 Aggregate	9.5%	6.4	4.8	3.9	3.4
1975 Aggregate	20.1	14.5	12.2	11.3	9.2
1969 Men	9.8	6.4	5.0	4.4	2.9
1969 Women	9.3	6.4	4.6	3.6	4.3
1975 Men	21.1	16.1	15.2	14.2	9.0
1975 Women	19.3	13.5	10.5	9.6	9.5
1969 Nonblack	9.1	6.0	4.6	3.6	3.1
1969 Black	11.2	7.9	5.3	5.4	6.4
1975 Nonblack	19.4	14.0	11.8	10.7	8.5
1975 Black	23.7	16.9	14.6	15.1	15.8
1975 Ages 16–24	31.7	21.3	17.7	16.1	17.2
1975 Ages 25–34	20.7	14.1	11.5	12.9	8.6
1975 Ages 35–44	12.7	10.8	9.6	7.5	6.4
1975 Ages 45–54	11.2	7.3	9.9	6.8	5.9
1975 Ages 55–64	4.8	9.9	6.1	7.3	5.5
1975 Ages 65+	5.5	10.5	6.7	9.1	4.9

Source: March Current Population Survey, Clogg & Sullivan, 1983, p. 145, Table IX. The actual sample size for the total labor force varies from 60,000 to 75,000. Total employment rates can be compared with published figures in the U.S. Bureau of Labor Statistics (1969, 1975), *Employment and Earnings*, April Issue.

lies for the ages 45 to 64, but in general the results suggest that [1.25 × PT] is a reasonable income cutoff.) The uniform standard of 1.25 × PT is low enough to isolate deficiency in earnings and to also isolate a fraction of the labor force with excessively high rates of unemployment.

Appendix Table 3.4 presents the unemployment by income contingency table that results from application of the 1.25 × PT rule to 1970 data, the only year for which comparable data are available from both Sullivan (1978) and Clogg (1979).

Appendix Table 3.4
Unemployment by Income, for Two Definitions
of the Low-Income Group, for 1970

| | Sullivan (1978) and Clogg (1979) | | New approach[b] | |
| | Income | | Income | |
	Low	Adequate	Low	Adequate
Unemployed	17.4%	3.5%	10.3%	4.0%
Employed	82.6%	96.5%	89.7	96.0
	100.0	100.0	100.0	100.0
% in class	9.4%	90.6%	9.4%	90.6%
Odds-ratio	5.8		2.8	
Log-odds–ratio	1.8		1.0	

[a]See Clogg (1979a, p. 88, Table 5.3), Clogg & Sullivan, 1983, p. 147, Table X. Percentages refer to people 14+ years old in the labor force.
[b]Low-income category is defined in relation to a standard of 1.25 * poverty threshold for an individual male in an urban area under the age of 65. Percentages refer to those 16+ years old in the labor force.

Unemployment risk is predicted better with the old approach, but the adequate income labor force has nearly identical unemployment risk with each method. The log-odds ratio, which summarizes the association between income and unemployment, is 1.0 with the new approach and 1.8 with the old approach. There is some slippage in the new definition of the low-income component (I), if we use the goal of maximizing association with unemployment as the sole criterion. But the approach nevertheless seems reasonable because it is simple to apply, it yields nearly identical measures of the incidence of the low-income form of underemployment, and it does not vary among individuals according to household characteristics (family size, primary earner versus secondary earner, and so on). Given the benefits associated with the new method, the decline in unemployment-low income association seems acceptable.

As a further test of the new method, we cross-classified unemployment (U), low hours (H), low income (I), educational mismatch (M), and adequate employment (A) using both the new and old approaches. (Because the subemployment component refers to people who are not currently in the labor force, it is excluded.) Appendix Table 3.5 has 12 cells rather than 16 because some combinations of the variable categories cannot occur together (e.g., unemployment and low hours, part-time employment).

Appendix Table 3.5
Cross-Classification of the Labor Force by Underemployment Form, 1970

LUF component U H I M	Sullivan (1978) and Clogg (1979) procedure[a]	New procedure	Difference[b]
+ – + +	0.1%	0.1%	0.0
– + + +	0.1	0.1	0.0
– – + +	0.6	0.7	–.01
+ – – +	0.3	0.3	0.0
– + – +	0.1	0.1	0.0
– – – +	9.4	8.0	1.4
+ – + –	1.5	0.9	0.6
– + + –	0.8	0.8	0.0
– – + –	6.3	6.8	–0.5
+ – – –	2.9	3.3	–0.4
– + – –	1.7	1.6	0.1
– – – –	76.2	77.3	–1.1
Total	100.0	100.0	0.0

[a]See Clogg (1979, p. 86)
[b]Index of dissimilarity between the two distributions is 2.1%. If calculations are based only on the percent distributions of the underemployed, the index of dissimilarity is 6.9%.
Source: Clogg & Sullivan, 1983, p. 147, Table XI.

Appendix Table 3.5 is noteworthy for the close similarity of the distributions from the two approaches. We regard the index of dissimilarity of 2.1% as acceptably small. Thus, the actual structure of underemployment inferred from the two approaches is very similar. The scaling and latent structure models applied by Clogg (1979, pp. 81–118) would lead to very similar results when applied to results from the new approach. Other studies of the comparability between the new approach and the old approach, including several other definitions of key LUF components, support the adoption of the simplified approach.

The measurement of mismatch follows exactly the techniques of Sullivan (1978) and Clogg (1979) and we present Appendix 3.4 for readers interested in replicating the measure. Labor force surveys in the United States since 1971 have used the 1970 Census classification of occupations; the 1980 Census classification scheme, which is quite different from the 1970 scheme, is not be used in standard surveys (like the CPS) until 1983. Mismatch criteria developed for the 1970 benchmark, using the 1970 Census classification, will therefore be useful for a wide variety of research purposes. Appendix 3.4 presents SPSS control cards that determine mismatch cutoffs. The REVED variable is completed years of schooling; it is constructed from the standard schooling variable (EDUC), which codes the highest grade of school that the individual has attended, and the variable COMGRD which measures (1 = yes) whether the individual completed the grade. NOCC is the three-digit occupation classification, and the statements break up the recoding into three separate tasks. A comment card toward the end of the figure indicates the special way that voluntary part-time workers and workers who do not have an occupation to report are handled. The reader may consult Sullivan (1978) or Clogg (1979) for further restrictions that might be of interest in measuring mismatch (e.g., the treatment of college students). Empirically, these definitions record mismatch only for individuals who have completed more than 12 years of schooling. Thus, the mismatch rates in this chapter are actually based on people with some post-high school education.

APPENDIX 3.4. SPSS CONTROL CARDS FOR DETERMINING EDUCATIONAL MISMATCH: 1970 CENSUS OCCUPATION CODES.

```
COMPUTE    REVED=EDUC
IF         (COMGRD EQ 1) REVED=EDUC+1
COMMENT ********************************************************************
           THE FOLLOWING MATCHING RECODE COMBINES
           OCCUPATIONS WHICH ARE HOMOGENEOUS FOR
           MEDIAN EDUCATION, BY MEAN YEARS OF EDUCATION
           PLUS ONE STANDARD DEVIATION.
           THIS MAPPING IS BASED ON A PREVIOUS ANALYSIS.
********************************************************************
COMPUTE    MISSA=NOCC
COMPUTE    MISSB=NOCC
COMPUTE    MISSC=NOCC
RECODE     MISSA(690=1)(390 820 831=2)(304 490 515 801 THRU 804 823
           824 834 902 970 972=3)(401 THRU 403 405 THRU 413 415 425
           531 THRU 444 452 472 475 THRU 480 491 492 495 THRU 501
           505 THRU 510 513 514 521 THRU 523 535 THRU 601 610 613
           621 630 THRU 640 642 THRU 645 651 THRU 673 675 THRU
           685 691 THRU 695 703 THRU 713 715 THRU 721 810 THRU
           814 821 830 835 841 851 854 874 THRU 901 903 960 THRU
           965 971 973 THRU 985=4)(ELSE=0)
RECODE     MISSB (265 303 305 THRU 320 323 THRU 360 381 THRU 383
           404 414 420 THRU 424 430 450 451 453 THRU 461 465 THRU 471
           473 474 493 494 502 THRU 504 512 520 524 THRU 530 602 THRU
           605 610 612 614 THRU 620 641 650 701 714 775 815 825 832 840 842
           THRU 850 852 853 860 905=5)(70 73 74 103 104 120 150 151 163 164
           181 185 THRU 192 250 THRU 262 275 THRU 280 290 301 302 370
           385 393 394=7)(10 THRU 12 14 15 20 72 84 101 154 161 165
```

```
                THRU 170 184 195 222 270 285 321 380=8)(ELSE=0)
RECODE          MISSC(0 13 21 22 75 THRU 83 85 THRU 93 102 111 130 160
                171 172 174 182 193 395=9)(23 THRU 60 71 105 131 THRU 145
                152 153 162 173 175 THRU
180
                183 184 200=10)(ELSE=0)
COMPUTE         MATCH=(MISSA+MISSB+MISSC)
COMMENT         **************************************************
                VOLUNTARY PART-TIME WORKERS (REASON35=12,13 OR
          EMPSTAT=5)
                AND THOSE WITHOUT AN OCCUPATION CODE
          (NOCC=999) ARE COUNTED AS NOT MISMATCHED.
**************************************************
COMPUTE         MISMATCH=2
IF              ((LABFORCE EQ 2) AND ((REVED-MATCH) GT 10))
                MISMATCH=1
IF              ((NOCC EQ 999) OR (LABFORCE EQ 2 AND ((REASON35
                GE 12 AND LE
          13)
                OR EMPSTAT EQ 5)))MISMATCH=2
IF              (LABFORCE EQ 0)MISMATCH=0
```

APPENDIX 3.5. A RATIONALE FOR COHORT ANALYSIS OF LABOR FORCE PARTICIPATION

Cohort analysts have been careful to note that age, period, and cohort are merely indicators of other variables that actually "cause" the observed variation in the dependent variable under study. The age–period–cohort framework is properly interpreted as an accounting scheme, not a "causal model," but we hasten to add that the proper application of this framework enables a deeper analysis of proximate causal mechanisms. Our view is that the selection of the proper causal variables to be considered in a modeling procedure is a most difficult theoretical task, one that is at least as difficult as applying the age–period–cohort accounting framework.

The age variable is biological, but aging is closely related to experience and schooling, the two most important factors that determine "offered wage" rates for prospective workers, and hence participation tendencies, according to the human capital theory (Mincer, 1962). Aging is also related to various status-role transitions (e.g., marriage, childrearing, retirement) that accompany a cohort's progression through an inherently age-graded life cycle (see Winsborough, 1978). Age differentials in participation (net of period and cohort) are probably not produced to any great extent by purely biological maturation. But once this much has been acknowledged, age differentials can be explained in causal terms only by properly specifying and measuring the life cycle events that account for age variability in a more precise causal sense. A method that incorporates the age variable explicitly into a model allows the determination of the combined influences of all the variables for which age is a proxy. But a method that incorporates the more proximate causal variables instead of the age variable must first be justified by carefully considering the proper specification of these variables, their functional relationship to the response, and their operational measurement.

The period variable is merely composed of observational units (usually years), but there are various meanings that are associated with it in the context of labor force participation. The effects of the period variable (net of age and cohort) could be due to the labor market variables (measured currently or in some specified lag scheme), which cause transitory "shocks" in labor force participation. But these period effects could be due to social movements that are only indirectly related to labor market conditions; the women's movement is certainly a period-specific phenomenon that could be relevant. Farkas (1977) and O. D. Duncan (1979) are led to conceive of period effects as those that are closely associated with certain (current) indicators of the condition of the labor market. But the decision regarding which indicators should be used poses a vexing problem, and we can never be certain that all of the period effect is captured by the set of indicators used. That these researchers conceive of period effects as only those that are related to the (current) economic condition of the labor market, excluding, say, period-specific social movements like the women's movement, is a serious problem.

The cohort variable is likewise one whose effects on participation can be viewed as a composite of more proximate causal effects. Cohort size (Easterlin, 1978, 1980; Welch, 1979), cohort-specific socialization influences, peculiar cohort experiences (the Depression, the Vietnam War), and human capital endowments (Weiss & Lillard, 1978) are only a few of the cohort attributes that might be relevant (see also Ryder, 1965, 1969; Glenn, 1977). Heckman's (1974) analysis of the determinants of the reservation wage of time (asking wages) for nonparticipants also reveals a possible role for cohort-specific causal mechanisms. Of the several determinants of reservation wages that he considers, the possible hours that a nonparticipant might supply to the work force, the wages of those on whom the nonparticipant depends, and household asset income might all be conceived at least partly as cohort-specific causal agents.

Because of the difficulties in specifying what it is that the age, period, and cohort variables actually represent in terms of proximate causal agents, an analysis that makes as few assumptions as possible about the causal agents is certainly legitimate.

APPENDIX 3.6. AGE-PERIOD-COHORT MODELS

The Main Effects Model

The models used to estimate the effects of age, period, and cohort on labor force participation are analogous to those used in a similar context by Clogg (1979). They are identical in most respects to those discussed by Pullum (1977, 1980) and Fienberg and Mason (1978), and these in turn are related to the diagonals parameter model of Goodman (1972). We let Ω_{ijk} denote the expected odds that a member of the i-th age, j-th period, and k-th cohort will be participating in the labor force, for $i = 1, \ldots, I; j = 1, \ldots, J; k = 1, \ldots, K$ (with $K = I + J - 1$, the number of cohorts in a cross-classification with I age categories and J periods). For the situation here all indices refer to single years, and the cohort indices range from the youngest to the oldest cohort. There is a linear relationship between the indices of the form $k = i - j + J$, and this accounts for the identification problem encountered in a model with main effects of the three indexing variables on the Ω_{ijk}.

The model that we consider first is

$$\Omega_{ijk} = \alpha \beta_i \gamma_j \delta_k \tag{3.1}$$

where α is a constant and β_i, γ_j, and δ_k denote age, period, and cohort parameters, respectively. There are $1 + I + J + K$ parameters in Eq. (3.1), and four (not three) restrictions must be imposed on them in order to achieve identifiability. For any given set of (unidentifiable) parameters in Eq. (3.1), define

$$f = (\delta_k / \delta_{k'})^{(k - k')^{-1}} \tag{3.2}$$

for some cohort category k^1 / K and some positive constant D. The parameters in Eq. (3.1) are then transformed by taking

$$\bar{\alpha} = \alpha \beta_I \gamma_J \delta_K f^{J-1} \tag{3.3a}$$

$$\bar{\beta}_i = (\beta_i / \beta_j) f^{i-1} \tag{3.3b}$$

$$\bar{\gamma}_j = (\gamma_j / \gamma_j) f^{j-J} \tag{3.3c}$$

and

$$\bar{\delta}_k = (\delta_k / \delta_K) f^{k-K} \tag{3.3d}$$

It can be verified that

$$\Omega_{ijk} = \bar{\alpha} \bar{\beta}_i \bar{\gamma}_j \bar{\delta}_k \tag{3.4}$$

using the relationship $k = i - j + J$, and that four independent restrictions have been imposed:

$$\overline{\beta}_I = \overline{\gamma}_J = \overline{\delta}_K = 1 \qquad\qquad (3.5a)$$

and

$$\overline{\delta}_k = \Delta \qquad\qquad (3.5b)$$

In addition to the three trivial restrictions of Eq. (3.5a) that serve to define the effects of age, period, and cohort category on the Ω_{ijk}, the k-th cohort effect has been set equal to Δ. Pullum (1980) actually considers situations where $\Delta = 1$, i.e., where the effect of the k-th cohort is assumed to be equal to that of the K-th cohort; Fienberg and Mason (1978) impose similar kinds of equality restrictions. But in view of these algebraic relationships, it is not necessary to assume equality of effects. Using algebraic identities like these, it is not difficult to derive the biases in the parameters that result from making an error of any specified amount in choosing Δ, a fact that will be exploited later. The reader is referred to Clogg (1982b) for algebraic details.

The necessary inputs to the model of Eq. (3.1) are thus two: the cohort category k' whose effect is to be restricted and the value Δ that is to be assigned to this effect. These specifications are very important. It can be shown that an error in choosing Δ produces systematic bias in all of the effects in the model, that this bias differs according to the cohort category k' chosen (with the bias being more severe when k' is not close to K, for a given error in specifying Δ). This bias depends as well on the number of age categories and the number of period categories used (a "design" effect). Therefore, special care must be taken to choose k' and Δ. Here we choose $k' = K - 1$ (the next-to-oldest cohort), and we choose in a manner that reflects a priori assumptions about intercohort trends in labor force participation. To assist in choosing Δ, note that if Ω_{ijk} is the value of the odds for the oldest cohort in any given age i and period j, then $\Omega^*_{ijk} = \Delta\Omega_{ijk}$ is the "pseudo-odds" that describes the expected odds, under the model, if cohort k' could actually be observed for age i and period j. The ratio of the second quantity to the first is Δ and so prior assumptions about the relative magnitude of these odds can be used to obtain a plausible value for Δ.

A Model with AGE–PERIOD Interaction

We assume that there exist ages i' and i'' such that for $i' < i < i''$ no age-period interaction exists. For our analysis we actually take $i' = 34$ and $i'' = 50$, the assumption being that no age–period interaction exists for the prime working ages 35–49. The choice of the ages is warranted, in our view, because the

35–49 age range is precisely that where unemployment is the lowest, where job security is the greatest, and where discouraged workers are virtually non-existent. Workers in these ages are typically those with much experience and their job skills are not yet outmoded. On the other hand, labor force participation is still in the process of being established for many people under age 35, and it is in the process of being terminated for many people over age 49. The assumption is then that the transitory period shocks on participation will exert their influences in different ways for each of the three age groups (20–34, 35–49, 50–64).

The model can be described by

$$\Omega_{ijk} \begin{cases} \alpha\beta_i\gamma_j\delta_k\,(1/\theta_{1j})^{i-c} & \text{for ages 20–34} \\ \alpha\beta_i\gamma_j\delta_k & \text{for ages 35–49} \\ \alpha\beta_i\gamma_j\delta_k\theta_{2j}^{i-c} & \text{for ages 50–64} \end{cases} \tag{3.6}$$

where c is a constant (equal to 23, the age index that corresponds to age 42). The thetas refer to age–period interactions for period j, for young and old workers. The age index i appears as a power of the thetas, incorporating the added assumption that there is a geometric relationship between the period shocks and age. Note that $\theta_1 > 1$ implies that the period shock operates to increase participation of younger persons relative to middle-aged persons, with a stronger relationship being exerted for the youngest persons; with $\theta_1 < 1$ the period shock operates to decrease the participation of younger persons relative to middle-aged persons. Similar comments apply to the θ_2 parameters governing the period shocks on older workers. The shocks are the greatest for the extreme ends of the ages 20–64, and they taper off to nothing in the center of the distribution. By virtue of the fact that there is no age–period interaction for ages 35–49, the model of (3.6) is identifiable, once the necessary restrictions have been imposed on the main-effects parameters.

Appendix Table 3.6 presents the degrees of freedom and indices of fit for models applied to gender and racial groups. The indices of dissimilarity provide the data for Figure 3.6 in the text.

Appendix Table 3.6
Indices of Fit

Model	df	Nonblack males		Nonblack females		Black males		Black females	
		L^2	D	L^2	D	L^2	D	L^2	D
Null	494	38,766	6.73%	16.380	6.62%	3793	7.76%	2435	7.31%
A	450	1,909	1.40	4,861	3.63	1032	3.83	835	4.47
P	484	38,328	6.70	13,738	5.73	3576	7.51	2321	7.05
A,P	440	1,489	1.20	2,106	2.40	811	3.40	730	4.09
C	440	13,967	3.85	5,477	3.77	1680	5.12	1208	5.22
A,C	396	764	0.81	609	1.33	776	8.30	587	3.70
P,C	430	12,800	3.46	4,933	3.49	1347	4.43	1121	4.90
A,P,AP*	418	708	0.79	921	1.65	769	3.31	653	3.85
A,P,C	387	508	0.710	519	1.21	645	2.96	541	3.48
A,P,C,CP*	365	509	0.66	471	1.17	620	2.91	517	3.38

Source: Clogg, 1982, **Demography**, p. 466, Table 1.

APPENDIX 3.7. CHOOSING IDENTIFYING RESTRICTIONS

The procedure used to identify the parameters is now outlined in some detail, bringing to light the prior information that we thought should be incorporated into the analysis. (Actually, for the case of nonblack males we concluded after the analysis that our prior assumptions were questionable, but we air them here nevertheless.) For each gender–race group, the 54th cohort (aged 63 in 1969) was singled out for the purpose of imposing the identifying restriction; the 54th cohort effect is restricted in relationship to the 55th cohort (aged 64 in 1969).

First consider nonblack men. Our prior assumption about them is that more recent cohorts will have lesser intrinsic tendencies to participate than older cohorts. Such a line of reasoning is certainly consistent with the observed period change for this group. The assumption made here seems to go hand in hand with the accepted notion that younger cohorts of nonblack women have greater intrinsic tendencies to participate and the labor force cannot expand without bounds; therefore, an adjustment in nonblack male cohort tendencies of this kind seems worth discussing. The value of Δ assigned to the 54th cohort effect is 0.98: The assumption is that the participation odds for the 54th cohort is 98% that of the 55th cohort. To see what this figure implies for labor force participation rates, suppose that $p = 0.90$ is the intrinsic labor force participation rate for the 55th cohort. (This value is approximately the average participation rate for nonblack men over the interval 1969–1979, and all that we require is that this value be somewhere near the "true" labor force participation rate for this cohort.) Now with $\Delta = 0.98$, the rate $p\% = p + e$

for the 54th cohort is 0.18% smaller than that of the 55th cohort. For values of p in the neighborhood of 0 .90, $\Delta = 0.98$ corresponds to about a 0.2% decline in the participation rate. If intercohort change were constant, in 55 years there would be about a 10% drop in the participation rate (to about 80%).

For nonblack women, the necessary prior information is much more clear-cut: newer cohorts of women must have a greater intrinsic tendency to participate than older cohorts. Such a prior assumption is consistent with earlier research and theory. The value of Δ is set at 1.02, and if the oldest cohort's rate of participation is $p = .55$, then this value of Δ implies that the rate $p\% = p + e$ for the 54th cohort is 0.49% larger than p. If intercohort change were constant, then over 55 cohorts there would be a 27% increase in the participation rate (to an average of 82%, or approximately the figure that we earlier entertained as the ultimate rate for nonblack men). Suffice it to say that $\Delta = 1.02$ appears credible for nonblack women (and $\Delta = 1.00$ would not be credible because it would imply no change over time).

For black men, we believe that the weight of the evidence supports a view that newer cohorts have lesser intrinsic tendencies to participate than older cohorts (see Wilson, 1996). We choose $\Delta = 0.99$ to reflect these assumptions, and note that if $p = 0.85$ is the 55th cohort's intrinsic rate of participation, then the rate for the 54th cohort will be 0.13% less than this. If the intercohort change were constant over 55 cohorts, this would imply a 7% drop in labor force participation. For black women, nearly the same prior information used for the case of nonblack women can be applied, although the pace of change for black women is probably not as great as for nonblack women. We choose $\Delta = 1.01$, and note that if $p = 0.60$ is the 55th cohort's intrinsic rate of participation, then the rate for the 54th cohort is 0.24% greater than that of the 55th. Assuming constant intercohort change, the rate of participation would rise 13% over 55 cohorts.

APPENDIX 3.8. RATE ADJUSTMENT TECHNIQUES BASED ON THE LOG-LINEAR MODEL

Rate adjustment procedures to adjust rates for changes in labor force composition were used to produce Tables 3.3 and 3.4. Assume that there are C composition categories, G groups being compared (one of which might be a standard), and D categories of the dependent variable. In this paper C referred to age–sex category ($C = 52$ levels), G to period ($J = 6, 12,$ or 13), and D to underemployment categories as measured by the LUF ($K = 6$). A cross-classification of C, G, and D produces a three-way contingency table that can be viewed as the primary set of data in most rate adjustment procedures. The approach here is based on the log-linear model for this 3-way contingency table. The relevant multiplicative (or log-linear) model is

$$F_{ijk} = \eta \tau_i^C \tau_j^G \tau_k^D \tau_{ij}^{CG} \tau_{ik}^{CD} \tau_{jk}^{GD} \tau_{ijk}^{CGD} \tag{3.7}$$

where F_{ijk} is the expected frequency in cell (i, j, k). The τ^{CG} parameters represent composition-group interaction ("group differences in composition") in the three-way table, while the t^{CGD} represent three-factor interaction. "Average" composition–category effects on the dependent variable are measured by the τ^{CD}; the group differences in the dependent variable are measured by the τ^{GD}.

Crude rates can be expressed in terms of the frequencies as

$$r_{\cdot j(k)} = F_{jk} / F_{+j+} \tag{3.8}$$

where "+" denotes summation over the subscript it replaces; $r_{\cdot j(k)}$ is the crude rate of prevalence of the k-th category of the dependent variable in the j-th group. From Eqs. (3.7) and (3.8) one can tell that the crude rates $(r_{\cdot j(k)})$ depend on the τ^{CG}, as well as the other parameters in the model. To obtain rates that do not depend on composition–group interactions, we have to remove the effect of τ^{CG}. Purged frequencies are defined by

$$F_{ijk}^* = F_{ijk} / \tau_{ij}^{CG} \tag{3.9}$$

These are free of the C–G interaction, and the adjusted rates are given by

$$r_{\cdot j(k)}^* = F_{+jk}^* / F_{+j+}^* \tag{3.10}$$

Note that the role of Eqs. (3.9) and (3.10) is merely to summarize the effect of all parameters in the log-linear model except τ^{CG}. Comparisons of the $r_{\cdot j(k)}$ (the crude rates) and the $r^*_{\cdot j(k)}$ can be used to determine the effect of the τ^{CG} on observed patterns among the crude rates.

To adjust rates for both C–G and C–G–D (three-factor) interaction, purged frequencies are obtained as

$$F_{ijk}^{**} = F_{ijk} / \left(\tau_{ij}^{CG} \tau_{ijk}^{CGD} \right)$$
$$= F_{ijk}^* / \tau_{ijk}^{CGD} \tag{3.11}$$

The F_{ijk}^{**} are used to calculate rates, $r_{\cdot j(k)}^{**}$, that are adjusted for C–G and C–G–D interaction. The $r_{\cdot j(k)}^{**}$ and the $r_{\cdot j(k)}^*$ can be compared with each other in order to determine the effect of three-factor interaction on the rates.

This approach is based on an adjustment for composition-group interaction in the 3-way table, as the τ^{CG} in Eq. (3.7) are measures of "partial" association between C and G. A different approach is to adjust rates for the composition–group interaction in the two-way C × G table; this table is a *marginal* table obtained by collapsing over levels k of D in the three-way table (i.e., $F_{ij} = F_{ij+}$).

Let γ_{ij}^{CG} denote the composition–group interactions in the $C \times G$ table. Now consider purged frequencies

$$F_{ijk}' = F_{ijk}/\gamma_{ij}^{CG} \tag{3.12}$$

Let $r_{.j(k)}'$ denote the rates calculated from the F_{ijk}'. These are adjusted for the *marginal* C–G interactions and they can be compared with the crude rates $(r_{.j(k)})$ much as before, or they can be compared with the $r_{.j(k)}^*$ (to determine whether there is a difference between adjusting for marginal C–G interaction and adjusting for partial C–G interaction). Next, the effect of three-factor interaction can be determined by considering purged frequencies

$$\begin{aligned} F_{ijk}'' &= F_{ijk}/\left(\gamma_{ij}^{CG}\tau_{ijk}^{CGD}\right) \\ &= F_{ijk}'/\tau_{ijk}^{CGD} \end{aligned} \tag{3.13}$$

Rates calculated from the F_{ijk}'' (say $r_{.j(k)}''$) can be interpreted in a way analogous to the way that the $r_{.j(k)}^{**}$ defined from Eq. (3.11) are interpreted.

When a particular group is selected as a standard to which other groups are to be compared, the results above can be modified appropriately. Let s denote the group taken as the standard. To create frequencies purged of the composition–period interaction in the three-way table, Eq. (3.9) is replaced by

$$F_{ijk}^+ = F_{ijk}\,\tau_{is}^{CG}/\tau_{ij}^{CG} \tag{3.14}$$

which does not alter the frequencies in the standard group (where $j = s$). Summary group rates calculated from the F_{ijk}^+, say $r_{.j(k)}^+$, will be such that $r_{.s(k)}^+ = r_{.s(k)}$, as is appropriate if the s-th group is the standard. Straightforward generalizations of this idea can be used for each of the other types of rate adjustment discussed earlier.

All of these different rate adjustment procedures were applied to the data discussed in this chapter. The section on composition-adjusted rates used adjusted rates $r_{.j(k)}^*$ and $r_{.j(k)}^{**}$. In the next section rates were further adjusted based on Eq. (3.14). These methods are described more fully in Clogg (1978, 1979, Chapter 4, Appendix H). Comparisons between these methods and more conventional "direct" standardization methods are taken up in these references and also in Keppel (1981), Clogg (1982), and Clogg and Shockey (1984a). However, the particular rate adjustments given in Eqs. (3.11)–(3.14) were not discussed in any detail in the previous literature.

4

Latent Class Models in the Analysis of Social Mobility

INTRODUCTION

Researchers studying social stratification have long been interested in the relationship between the socioeconomic status of parents and children. They have focused on this relationship for two reasons. First, connections between the socioeconomic status or class of parents and children reveal differences in life chances that generally cannot be attributed to individual efforts of children themselves. Second, the precise assessment of the relationship between origin and destination socioeconomic statuses across generations reveals the location of significant departures from meritocracy. Virtually all advanced industrialized nations express some normative and ideological commitment to meritocracy—the idea that one's place in society should derive from one's own efforts and not reflect accidents of birth or excessive help from family networks. Indeed, the classical sociological statements on what differentiates modern from premodern social organization almost always emphasize the rise of meritocratic selection and the decline of family origins as an important organizing principle of stratification systems.

Latent structure models of mobility are one way that social scientists attempt to understand the relationship between social origins and destinations. To understand how these techniques work, it is important to understand what researchers are trying to accomplish and what information they have to work with.

Modeling Social Mobility—What Researchers Want to Know

Most students of social stratification are interested in studying the degree of meritocracy and openness in a society, and significant departures from these. Studying the strength of intergenerational ties in labor market processes is important because it helps us to understand the degree of meritocracy and openness in a society.

Here, we define a few terms for those unfamiliar with this type of research. *Social mobility* generally refers to individual movements in the stratification system, or to changes in an individual's position in the social hierarchy due to changes in the society's stratification system. *Intergenerational mobility* refers to changes in social status across generations from parents to children, or in some cases from grandparents to grandchildren. Such general questions as "will our children be better off than their parents?" are community and societal-level counterparts to the question researchers ask, "what is the relationship between parents' and children's social status?" *Meritocracy* refers to the organization of a stratification system based on the acquired skills and actions of individuals, rather than on ascribed attributes such as individual class locations, race, gender, national origin, or social origins.

When researchers study intergenerational mobility, they are looking for specific mechanisms that link the resources of parents to children and the pattern of those connections. Much of the analysis of intergenerational mobility occurs by examining mobility tables. Mobility tables are cross-classifications of origin by destination classifications, and typically describe the relationship of parents' socioeconomic or class position to that of their children. Examples of three mobility tables are presented in Table 4.1.

In this typical table, father's occupational status is given as the row variable and that of the son is given as the column variable. The numbers in the table tell us how many father–son pairs (in the sample) experience a specific origin–destination status combination. The entries on the diagonal (situations where fathers and sons have the same occupational status) typically have the highest frequency—many fathers and sons have the same social status. Also notice that there are few extreme movements in the table. Few sons of low status fathers end up in high social status jobs, and vice versa.

But researchers are often interested in more than the descriptive relationship between social origins and destinations. An important goal for many is to understand how parental socioeconomic status translates into advantages and disadvantages for children. Specifying the mechanisms that translate social origins into destinations is similar to producing "road maps" of how people raised by parents with certain characteristics and resources "get to" the places they occupy as adults in the stratification system.

Table 4.1
Three Mobility Tables for Denmark and Great Britain

A. Danish 5 × 5 Table (N = 2,391)

Father's Status	Subject's Status				
	1	2	3	4	5
1	18	17	16	4	2
2	24	105	109	59	21
3	23	84	289	217	95
4	8	49	175	348	198
5	6	8	69	201	246

B. British 5 × 5 Table (N = 3,497)

Father's Status	Subject's Status				
	1	2	3	4	5
1	50	45	8	18	8
2	28	174	84	154	55
3	11	78	110	223	96
4	14	150	185	741	447
5	0	42	72	320	411

C. British 8 × 8 Table (N = 3,497)

Father's Status	Subject Status							
	1	2	3	4	5	6	7	8
1	50	19	26	8	7	11	6	2
2	16	40	34	18	11	20	8	1
3	12	35	65	66	35	88	23	21
4	11	20	58	110	40	183	64	32
5	2	8	12	23	25	46	28	12
6	12	28	102	162	90	553	230	177
7	0	6	19	40	21	158	143	71
8	0	3	14	32	15	126	91	106

Taken from Clifford C. Clogg, 1981 "Latent Structure Models of Mobility." *American Journal of Sociology,* Vol. 86, p. 844.

Modeling Social Mobility—What Researchers Have to Work With

Unfortunately, it is producing that road map that presents significant difficulties to the researcher. Often all we have when attempting to understand social mobility is a set of origin and destination positions arrayed as a mobility table, like the one described above. The road map for linking the two is often completely inaccessible. That is, it cannot be observed by the researcher.

For example, let the origins and destinations in a mobility table be labeled O and D, respectively. Let the pathway(s) through which individuals must travel to get from O to D be labeled generically as X. Figure 4.1 presents a simple diagram representing this logic.

Our difficulty arises when the different types, or classes, of pathways and their attributes are not observable. In this setting we say that X is a latent variable or, more precisely, that X constitutes a set of latent classes (hence, "latent class" analysis). To see what X may represent in the context of social mobility, consider a fundamental empirical relationship found in almost all of the studies on social mobility in industrialized nations. This well-documented empirical fact is that there is a positive relationship between an individual's education and his or her social status. Moreover, education usually intervenes in the relationship between origin and destination statuses, making education one of the likely components making up our latent variable X. But remember, in the context of a mobility table of the type described above we

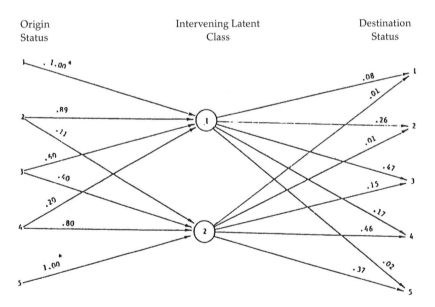

Figure 4.1. Example of two latent classes as intervening variables between social origins and destinations. Numbers are transition rates into and out of the intervening classes, taken from the Danish mobility table. Taken from Clifford C. Clogg, "Latent Structure Models of Mobility" 1981. *American Journal of Sociology*, Vol. 86, p. 850.

cannot actually observe X. Because of that, we would never be able to actually observe anything in X. But if we could, we would likely see that the variables making up X were empirically related to education.

Now suppose that, beyond education, there are even more mechanisms linking social origins and destinations whose conjunctions are characterized by their lack of observability. These may range from parents' and peers' orientations toward education to the attainment of status to network connections influencing job placement. There are, of course, many (unobserved and unobservable) influences governing the relationship between origins and destinations. The situation is complicated further because the likely benefits or decrements that accrue to children who possess or lack parental resources probably depend on the position from which one starts.

As these factors grow in number, and the conjunctions between these factors grow in complexity (as our roadmap becomes cluttered with an ever-growing set of connections), it becomes increasingly difficult to properly specify the precise observed mechanisms governing the path from destinations to origins. Nevertheless, we can still infer the minimally sufficient number of unique sets of pathways necessary to explain the links between origins and destinations, and the character of each pathway set in terms of origin and destination probability distributions. That is, we can focus in on the number of pathway classes (the latent classes) that bind, in a probabilistic manner, specific origins to specific destinations. Further, we can link origin/destination combinations to specific latent pathway classes. Latent class analysis helps us reveal the number and composition of these latent classes (or classes of pathways) that explain the relationship between origins and destinations. The latent classes can be considered to house the predominant pathways binding origins and destinations.

THE DATA

Table 4.1 presents three mobility tables pertaining to intergenerational occupational mobility in Denmark and Great Britain (Glass, 1954; Svalastoga, 1959; Miller, 1960). These tables, or various condensed versions of them, have been analyzed so extensively that little comment about their makeup is required.

The rows and columns in the 8 × 8 British table correspond to the following occupational categories: (1) professional and high administrative; (2) managerial and executive; (3) inspectional, supervisory, and other nonmanual (high grade); (4) inspectional, supervisory, and other nonmanual (low grade); (5) routine grades of nonmanual; (6) skilled manual; (7) semiskilled manual; (8)

unskilled manual (Miller, 1960, pp. 70–72).[1] These data allow two distinct types of comparative analysis; a direct comparison of mobility structures in Denmark and Great Britain based on the two 5 × 5 tables and a comparison of mobility structures in a relatively coarse and in a relatively fine categorization of occupational categories for Great Britain.[2]

Two-Class Models

Table 4.2 presents several summary statistics and indices pertaining to unrestricted models of the three sets of mobility data. The usual chi-square statistics are reported, with chi-square referring to the Pearson goodness-of-fit statistic and L^2 to the chi-square statistic based on the likelihood-ratio criterion. The index of dissimilarity between the observed and expected frequencies is presented as a simple index of fit and will be used to compare models. We use it here because it does not depend in a direct way on sample size, so it can be used to make comparison between models applied to the Danish table (where $N = 2391$) and models applied to the British tables (where $N = 3497$).

Other indices presented in Table 4.2 allow assessment of the fit from different points of view. One is the likelihood-ratio chi-square statistic divided by its degrees of freedom, $F = L^2/df$, a quantity analogous to the F-statistic in regression analysis (see Haberman, 1978, p.17). A final index of fit is the R^2, which is calculated by using the L^2 of the baseline model of independence

Table 4.2
Unrestricted Latent Structure Models: Indices of Fit

Model	df	L^2	X^2	Δ (%)	$F = L^2/df$	R^2
			Danish 5 × 5 Table			
H_1	16	654.2	754.1	19.4	40.9	—
H_2	9	116.2	133.6	8.8	12.9	.82
H_3	2	32.7	33.3	5.5	16.9	.95
			British 5 × 5 Table			
H_1	16	811.0	1,199.4	15.4	50.7	—
H_2	9	182.3	206.7	8.0	20.3	.78
H_3	2	55.1	49.8	4.6	27.6	.93
			British 8 × 8 Table			
H_1	49	954.1	1,415.3	16.6	19.6	—
H_2	36	253.5	282.4	9.3	7.0	.73
H_3	21	96.7	99.3	6.5	4.6	.90

Taken from Clifford C. Clogg, 1981, "Latent Structure Models of Mobility." *American Journal of Sociology*, Vol. 86, p. 846.

(namely, the model of perfect mobility) as a measure of overall variability to be accounted for and the L^2 of successive models as measures of residual variability.[3]

Model H_1 in Table 4.2 gauges whether one underlying relationship exists between social origins and destinations. This model and the summary indices are presented for the purposes of gauging the improvements produced by more complicated models. Using the pathway imagery described above, this one-class model posits that any pathway from any origin to any destination is taken merely by chance. This model posits no structure in the relationship between origins and destinations.

The two-class latent structure model denoted by H_2 provides a moderate improvement over H_1 for each set of data. Substantively, this model posits that there are two primary sets of pathways structuring the relationship between origins to destinations. The index of dissimilarity, Δ, ranges from 8.0% to 9.3%, and the R^2 ranges from 0.73 to 0.82.

For reasons described in Appendix 4.1, we need to make two assumptions about the character of the two latent classes to properly identify parameter estimates in the two-class model. One useful assumption for distinguishing the two latent classes is to prevent those from the lowest origin status (level 5) from membership in the first latent class. This is equivalent to stating that the pathways linking origins to destinations in the first latent class cannot include a connection to the lowest origin status, but only connections to the top four origin statuses.

Similarly, we assume that the second latent class does not contain members from the highest origin status, origin level 1. This is equivalent to stating that the pathways linking origins to destinations in the second latent class cannot include a connection to the highest origin status, but rather only connections to the bottom four origin statuses. Combining these two assumptions, latent class one is characterized as the upper latent class and latent class two is characterized as the lower latent class, in terms of the starting points on the pathways.

Given these identifying restrictions, Panel B of Table 4.3 displays the estimated conditional probabilities of observing origins (upper set of probabilities) and destinations (lower set of probabilities) given each latent class. There are several characteristics of these results that are worth noting:

1. The probability of being in latent class one (the upper class) is greater than the probability of being in latent class two (the lower class) for those who come from higher social origin categories (categories 1, 2, and 3).

2. The probability of being in the second, lower latent class is greater than the probability of being in the higher latent class for those in the bottom social origin categories (categories 4 and 5).

Table 4.3
**Different Parameter Estimates for the Two-Class Model Applied
to Danish Data**

	A Latent Class t		B Latent Class t	
	$t = 1$	$t = 2$	$t = 1$	$t = 2$
$\hat{\pi}_{it}^{\bar{O}X}$:				
$i = 1$	.0619	.0016	.0620	.0000*
2	.3080	.0305	.3083	.0234
3	.4616	.1992	.4619	.1924
4	.1681	.4175	.1678	.4239
5	.0004	.3512	.0000*	.3603
$\hat{\pi}_{jt}^{\bar{D}X}$:				
$j = 1$	.0785	.0065	.0755	.0065
2	.2753	.0132	.2645	.0134
3	.4876	.1508	.4738	.1510
4	.1545	.4593	.1670	.4591
5	.0042	.3702	.0192	.3700
$\hat{\pi}_t^X$:	.3693	.6307	.3847	.6153
	$L^2 = 116.232$		$L^2 = 1216.232$	

*Denotes a restriction; see text for details.
Taken from Clifford C. Clogg, 1981, "Latent Structure Models of Mobility." *American Journal of Sociology,* Vol. 86, p. 847.

3. 38% of the Danish sample are in the upper class, and 62% are in the lower class.

Although the two-class model does not fit the data well, it is still instructive to consider details here as it is among the simplest of our models. Table 4.4 displays the assignments of different origin/destination combinations of the Danish mobility table to latent classes, and summarizes the error rates of these assignments. The error rate is the probability that a person assigned to a cell is drawn from a latent class other than the one assigned. An error probability of zero indicates that that cell is assigned to the corresponding latent class with certainty, while anything greater than zero indicates the degree of uncertainty in the assignment.

Overall, 88% of the respondents are correctly allocated to latent classes using this model, with a lambda measure of 0.693. Also notice that the error rates are greatest toward the middle of the table and smaller as we move toward the corners. The assignments are most uncertain for transitions from origin status 4 (routine grades of nonmanual occupations and skilled manual occupations) to destination status 3 (low-grade nonmanual occupations), and for transitions from origin status 3 to destination status 4. This indicates that

Table 4.4
Assigning Cells from the Danish Mobility Table to Latent Classes (Using the Two-Class Model with Identifiable Parameter Estimates)

	D = 1	2	3	4	5
O = 1	1	1	1	1	1
	.00	.00	.00	.00	.00
2	1	1	1	1	2
	.01	.01	.04	.25	.30
3	1	1	1	2	2
	.06	.03	.16	.35	.07
4	1	1	2	2	2
	.26	.17	.44	.08	.01
5	2	2	2	2	2
	.00	.00	.00	.00	.00

% correctly allocated into latent classes = 88.2%.
Lambda = .693.
Note – The first entry in each cell is the class assignment under the 2-class model with $\pi_{51}^{OX} = \pi_{12}^{OX} = 0$. Class 1 is "upper"; class 2 is "lower". The second entry is the probability that a person with the given mobility pattern is drawn from a laten class different from the one assigned under the model.
Taken from Clifford C. Clogg, 1981, "Latent Structure Models of Mobility." *American Journal of Sociology*, Vol. 86, p. 849.

we have the least amount of uncertainty for the latent class assignments involving these origin/destination combinations. Using the pathway imagery described above, this indicates that, although these origin/destination combinations most often follow the latent pathways describing the lower class (latent class two), they nevertheless often behave as if they follow intervening pathways describing the upper class (latent class one).

Three-Class Models

Model H_3 in Table 4.2 fits a three-class model to the British and Danish data. This model tests for the existence of an upper, middle and lower class in our mobility tables. This model is an improvement over the first two but still does not fit the data well.[4] An examination of the residuals from these models reveals that, not surprisingly, the failure to fit the data is largely due to a lack of fit on the diagonals. That is, these unrestricted latent class models fail to adequately account for the inordinate tendency for fathers and sons to have the same occupational classification. We therefore turn to different types of restricted latent class models to account for the high level of nonmovers in the mobility table.

SOME RESTRICTED LATENT STRUCTURES

In this section we examine models with more restrictions in an attempt to highlight the underlying patterns of mobility and immobility. This section applies some restricted latent structures to the data with the objective of testing the models and arriving at the substantive implications of them for the study of mobility. All of these models are extensions of the one- and two-class models we have just discussed.

The first model we consider tests for the existence of "latent stayers" and "latent movers," and is equivalent to the quasi-perfect mobility model considered by Goodman (1969, 1972). For this latent mover–stayer model, as it is sometimes called, there are as many latent stayer classes as there are origins. For each origin, there is a deterministic class of "stayers," those whose desti-

Table 4.5
Restricted Latent Structure Models: Indices of Fit

	df	L^2	X^2	Δ (%)	$F = L^2/df$	R^2
			Danish 5 × 5 Table			
H_1	16	654.2	754.1	19.4	40.9	—
M_1	11	248.7	270.3	9.5	22.6	.62
M_2	7	43.2	43.0	4.5	6.2	.93
M_3	6	8.2	8.1	1.7	1.4	.99
M_4	5	8.2	8.1	1.7	1.6	.99
M_5	4	8.2	8.1	1.7	1.6	.99
M_6	7	27.9	33.4	2.6	4.0	.96
			British 5 × 5 Table			
H_1	16	811.0	1,199.4	15.4	50.7	—
M_1	11	249.5	327.3	6.6	22.7	.69
M_2	7	33.2	33.8	2.5	4.7	.96
M_3	6	12.9	12.5	1.7	2.2	.98
M_4	5	12.9	12.5	1.7	2.6	.98
M_5	4	12.9	12.5	1.7	3.2	.98
M_6	7	71.4	76.2	3.7	10.2	.91
			British 8 × 8 Table			
H_1	49	954.1	1,415.3	16.6	19.5	—
M_1	41	446.8	55.1	10.1	10.9	.53
M'_4	29	60.3	61.1	4.1	2.1	.94
M'_5	28	60.3	61.1	4.1	2.2	.94

Taken from Clifford C. Clogg, 1981, "Latent Structure Models of Mobility." *American Journal of Sociology*, Vol. 86, p. 854.

nation is the same as their origin. For a so-called stayer, there is a direct "inheritance pathway" between the same origin and destination, a pathway taken with certainty. Within the latent mover class, mobility is perfectly random. Mobility across statuses is prohibited (i.e. stayers cannot become movers, and vice versa). This model is designated M_1 in Table 4.5, and we see clearly that it does not fit any of the sets of data (the deltas, e.g., range from 6.6% to 10.1%).

A Two-Class Quasi-Latent Structure: Indices of Fit

Let us consider a two-class latent structure in which two classes of latent movers (namely, an upper and a lower class) are posited to exist. In addition, let us posit the existence of K stayer classes corresponding to K ($\leq$ I) of the statuses in the table. Such a model modifies H_2 in a way analogous to the way in which M_1 (the quasi-perfect mobility model) modifies H_1. From a purely statistical point of view, these models are constructed to account for the clustering on the main diagonal, and this is done by introducing the concept of deterministic status inheritance classes.[5]

Models M_2–M_6 introduce deterministic stayer classes into the two-class models for the 5 × 5 table, and these greatly improve the fit. Model M_2 introduces two status classes, one for each of the statuses 1 and 5; there are accordingly two fewer degrees of freedom for M_2 relative to H_2. Model M_3 introduces a third status class corresponding to latent stayers in status 3 and produces a most compelling fit for both sets of data.[6] Note that Δ is 1.7% for M_3 applied to each set of data, and $R^2 = 0.99$, 0.98, $F = 1.4$, 2.2, respectively, for the Danish and British tables.[7]

Parameter Estimates from the Quasi-Latent Structure

The parameter estimates associated with M_3 for the two 5 × 5 tables are presented in Table 4.6. The estimated proportions in the latent upper class are 0.251 for the Danish data, 0.220 for the British data. After status inheritance is introduced by considering three stayer classes, the proportion in the upper class in the two societies is virtually identical. Similarly, the estimated proportions in the lower class are 0.624 for Denmark and 0.691 for Britain, implying that the British data exhibit a slightly heavier concentration in the lower class. What is striking about the differences between the two sets of data is the proportions in the three stayer classes. The Danish data have one-half the proportion of latent stayers in status 1 (0.006/0.012), slightly over four times the proportion of latent stayers in status 3 (0.059/0.014), and roughly equal numbers who are latent stayers in the fifth status. These results correspond to 25, 20, and 27% of those in origin statuses 1, 3, and 5 being latent stayers in Denmark, and 32, 10, and 26% in Britain. In both absolute and

Table 4.6
Different Parameter Estimates from Model 3, Table 4.5

	Danish 5 × 5	British 5 × 5
$\hat{\pi}^X$	.251	.220
$\hat{\pi}_2^X$	.624	.691
$\hat{\pi}_3^X$	.006	.012
$\hat{\pi}_4^X$	.059	.014
$\hat{\pi}_5^X$	.059	.062

B. Conditional Probability of Origin Status I, Given Latent Class t $(\hat{\pi}_{it}^{\bar{O}X})$

	Danish 5 × 5		British 5 × 5	
i	$t = 1$	$t = 2$	$t = 1$	$t = 2$
1	.071	.000+	.111	.000+
2	.450	.032	.509	.043
3	.342	.242	.178	.137
4	.137	.466	.202	.560
5	.000+	.260	.000+	.260

C. Conditional Probability of Destination Status j, Given Latent Class t $(\hat{\pi}_{jt}^{\bar{D}X})$

	Danish 5 × 5		British 5 × 5	
j	$t = 1$	$t = 2$	$t = 1$	$t = 2$
1	.084	.010	.077	.000
2	.383	.022	.438	.063
3	.375	.195	.165	.117
4	.125	.505	.254	.510
5	.032	.269	.067	.310

Taken from Clifford C. Clogg, 1981, "Latent Structure Models of Mobility." *American Journal of Sociology*, Vol. 86, p. 856.
+ denotes a restriction.

relative terms, the differences observed between the two sets of data with respect to the latent class distribution arise because of differing tendencies for status inheritance in the first and third statuses.

The Prediction of Membership in Latent Classes and Latent Status Classes

We shall now describe the latent class predictions for all three sets of data based on models that fit. The only difference between the predictions and those presented earlier is that now stayer classes must be taken into account.

Table 4.7 summarizes the predictions for the two 5 × 5 tables. The most salient feature of this comparison is the striking degree of similarity in results, as can be seen in several ways. First, the predictions are identical for each table, with the single exception of the prediction for the (4, 2) mobility pattern. With the Danish data (Table 4.7A) the (4,2) pattern of mobility is sufficient to produce a prediction of membership in the upper class (with an error rate of 0.33), while with the British data (Table 4.7B) this same pattern of mobility produces a prediction of lower-class membership (with an error rate

Table 4.7
Assigning Cells from the Danish and British Mobility
Tables to Latent Classes

		Destination Status				
	A.	1	2	3	4	5
Origin	1	3*	1	1	1	1
Status		.20	.00	.00	.00	.00
	2	1	1	1	1	2
		.02	.01	.18	.42	.40
	3	1	1	4*	2	2
		.17	.19	.51	.12	.06
	4	1	1	2	2	2
		.49	.33	.18	.13	.01
	5	2	2	2	2	5*
		.00	.00	.00	.00	.42

% correctly allocated into latent classes = 82.9.
Lambda = .545.

		Destination Status				
	B.	1	2	3	4	5
Origin	1	3*	1	1	1	1
Status		.13	.00	.00	.00	.00
	2	1	1	1	1	2
		.00	.04	.16	.35	.45
	3	1	1	4*	2	2
		.01	.26	.56	.17	.08
	4	1	2	2	2	2
		.03	.45	.14	.15	.02
	5	2	2	2	2	5*
		.00	.00	.00	.00	.47

% correctly allocated into latent classes = 83.7%.

Lambda = .472.
* Asterisked numbers denote a prediction of membership in a particular status -class.
Taken from Clifford C. Clogg, 1981, "Latent Structure Models of Mobility." *American Journal of Sociology*, Vol. 86, p. 863.

of 0.45). Interpreted in another way this says that 67% of the those with this pattern of mobility would be assigned to the upper class in Denmark, while only 45% would be assigned to the upper class in Britain. We remarked earlier that Britain appears to have a slightly heavier concentration in the lower class, and this characteristic shows up in the prediction for this particular mobility pattern. As a whole, the quality of the prediction is comparable for each table: 82.9 and 83.7% of respondents would be correctly allocated to the latent classes and stayer classes for Denmark and Britain. The upper- versus lower-class division in both tables is comparable, with the division occurring approximately on the diagonal of the table that runs from the lower left to the upper right.

Table 4.7 also provides us with the error rates associated with the predictions of membership in the three stayer classes associated with M_3. For example, the error rate of 0.20 for the (1,2) cell for the Danish table signifies that 80% of the observed stayers in status 1 can be regarded as latent stayers. Similarly, 87% of the stayers in status 1 for the British table can be regarded as latent stayers. As a direct consequence of the model, it is easy to see that all other members of the (1,1) mobility pattern besides those in the associated status class are members of the latent upper class, and this is a consequence of the fact that the restriction was applied that nobody from origin status 1 was allowed to be in the second (lower) latent class. Because this holds, it appears appropriate further to characterize the status class associated with the first origin status as one embedded in the upper class. To corroborate this interpretation, merely note that all mobility patterns adjacent to the (1,1) cell produce predictions of membership in the upper class, and note in addition that membership in the first origin status denotes membership in either the first stayer class or the upper class. Because of these considerations, it seems justified to speak of the first latent status class as a special category within the upper class, one whose members are especially influenced by inheritance of their upper class origins.

Similar comments apply directly to the characterization of the stayer class connected with the fifth status. We are led to characterize this group (denoted as class 5 in Table 4.7) as a stayer class within the lower class, and this designation is a consequence of the fact that all members of origin status 5 are predicted to be in either the lower class or that stayer class, and/or the fact that all mobility patterns adjacent to the (5,5) cell are assigned to the lower class. As with the case for the upper class, this stayer class is similarly influenced by inheritance, but with ties to lower class origins.

A similar interpretation cannot be made for the stayer class corresponding to the third observed status, as (1) those not in the relevant status class (denoted as class 4 in Table 4.7) might be in either the upper or the lower class, and (2) the mobility patterns adjacent to the (3,3) cell produce predictions of both upper- and lower-class membership, depending on the direction in which one moves from this cell. The status class corresponding to latent stayers in observed status 3 is thus intermediate between the upper and lower classes indirectly observed in the table.

In sum, there is an appealing simplicity underlying these rather complex empirical models and results. The pathways linking origins to destinations may be sufficiently characterized by three deterministic stayer classes and two probabilistic latent classes, or two primary sets of pathways, linking origins and destinations. Moreover, two of the three deterministic stayer classes (the upper and lower status classes described above) may be viewed as special cases of the two latent classes, cases in which inheritance is especially strong. Finally, the third stayer class straddles the two latent classes, taking the position of what may be construed as the middle class.

Returning to our pathway imagery discussed at the beginning of this chapter, we can think of these classes as constituting five distinct segments on the map linking origins and destinations. Two of the five segments, those constituted by upper class and lower class stayers, show strong connections, or well-worn paths if you will, linking similar origins and destinations. These two segments clearly show high levels of inheritance. Precisely how this inheritance operates we cannot know with this type of analysis. Nevertheless, the analysis does point us toward the fact that, indeed, upper and lower classes tend toward a significant degree of inheritance above and beyond that observed in the so-called middle classes. The remaining three latent classes are less influenced by inheritance, and are thus necessarily more influenced by other socioeconomic and societal factors.

CONCLUSION

We have tried to show in this chapter how the latent class model can be used to provide insights into the analysis of social mobility. These methods are consistent with a variety of important conceptual paradigms that have been developed for understanding social mobility. As we have shown here, the latent class model seeks to uncover the number, and to an extent the characteristics, of unique sets of pathways intervening between origins and destinations. We found that three classic sets of mobility data, which have been analyzed extensively over the past 20 years, could be adequately described by a fairly simple latent structure, consisting of movers and stayers, along with two latent classes defining upper and lower social classes. This model is consistent with Weber's idea that mobility within social class or status should be relatively easy, while mobility across social class or status boundaries should be more difficult.

NOTES

1. For the 7 × 7 table published in Glass, status categories 5 and 6 were combined; and in the 5 × 5 table given in Table 4.1, status categories 2 and 3, status categories 5 and 6, and status categories 7 and 9 were combined. Svalastoga (1959) is responsible for collapsing the British table into its 5 × 5 form; he did so to make it more comparable to the Danish 5 × 5 table reported in Table 4.1.

2. Special note should be taken of the zero count for the (5,1) cell in the British 5 × 5 table and of the zero counts in the (7,1) and (8,1) cells in the British 8 × 8 table. These zero counts will require simple modification of some of the latent structure models we use. The British 8 × 8 table presented by Miller (1960) and used by O. D. Duncan (1979) is apparently in error (see corrections described in Clogg, 1981).

3. Indices of fit not directly based on the magnitude of χ^2 or L^2 are required because the data have been obtained by rather complicated sampling arrangements, and the descriptive levels of significance associated with the χ^2 statistics should not be taken at face value.

4. Going with a three-class model is a natural choice when a two-class model does not fit the data. However, there are difficulties in the application of the three-class model to mobility tables. At least two restrictions are required, and when two restrictions of the kind used in connection with H_2 are imposed (see Table 4.3) the fit of the model is equivalent to that obtained for the completely unrestricted model, $L^2(H_3) = 32.7$, $\Delta = 5.0\%$. However, even with two restrictions, the parameter estimates are still unidentifiable, and two additional restrictions are required. The three-class model does improve the fit, but it is clearly not parsimonious. Moreover, it still does not fit the data to an acceptable degree as judged by any of the indices of fit.

5. For the 5 × 5 tables, the same kind of identification problem arises with these models as was encountered with H_2. Accordingly, the restrictions used above to identify parameter estimates will be employed here. Those from the highest origin status cannot end up in the lowest destination status, and those in the lowest origin status cannot end up in the highest destination status. The zero count in the (5,1) cell of the British table also has the same consequence for applying the methods and so this count was replaced by 0.1 to apply the models directly.

6. If the χ^2 statistics were taken as pertaining to simple random samples (i.e., multinomial samples) or to stratified random samples with the destination statuses as the stratifying criterion (i.e. product-multinomial samples), the descriptive level of significance for M_3 would be greater than 0.20 for the Danish table and only slightly less that 0.05 for the British table.

7. Duncan, O. D. (1979) obtained a somewhat better fit for the 8 × 8 table with his row-effects and his uniform-association model, with the diagonal "deleted" ($L^2 = 45.9$, 58.4 on 34 and 40 df, respectively), but his model is developed from an entirely different point of view than the models considered here (see also Goodman 1979). Duncan examined the dependence of destination status on origin status and formulated coefficients somewhat analogous to the simple regression coefficient β_{do} in the fashion of status attainment research (see Blau & Duncan, 1967).

Models M_4 and M_5 posit four and five status classes, respectively, and contribute nothing to fit. The estimated proportions of latent stayers in statuses 2 and 4 turn out to be nearly zero for each set of data, given that latent stayers are admitted for statuses 1, 3, and 5. Model M_6 considers the possibility that latent stayers in status 1 can be deleted from model M_3, but for each set of data it is clear that this is not tenable. Model M_3 thus appears as a plausible model for both 5 × 5 tables, and there appear to be no serious competitors involving either more or fewer status classes than the three considered with M_3.

We also considered similar models for the British 8 × 8 table, namely, models M = 4 and M = 5. Model M = 4 modifies the two-class model by positing the existence of seven status classes (a status class corresponding to status 3 was omitted). Model M = 5 is the model with all eight status classes considered, and we see that this does not improve the fit to any discernible extent.

Model M = 4 is thus the quasi-latent structure suggested for the 8 × 8 table. It fits the data less well than M_3 applied to the 5 × 5 tables and would still be judged inadequate if the χ^2 statistics were taken at face value. However, M = 4 accounts for 94% of the variability in the table, has a delta of 4.1% and an F of only 2.1, implying that the model performs very well indeed when compared with any of the other models considered in this paper. We have focused analysis directly on the class structure of mobility processes, taking the variable X as the intervening variable that explains the dependence of destination on origin. While the fit of M = 4 can certainly be improved, inspection of the R^2 or Δ indices indicates that only a little improvement is necessary.

APPENDIX 4.1. THE LATENT STRUCTURE APPROACH TO THE ANALYSIS OF MOBILITY TABLES

The Latent Structure Model

Let π_{ij} refer to the expected proportion in the (i,j) cell of an $I \times I$ mobility table, where $I = 1, \ldots, I$ and $j =, \ldots, I$. The row variable will refer to the origin statuses (e.g., father's occupational category) and will be denoted by the symbol O. The column variable will refer to the destination status (e.g., son's occupational category) and will be denoted by the symbol D. With $\pi_i^0 = \Sigma_j \pi_{ij}$ defining the marginal distribution of the statuses at the time of origin, the mobility rates can be written as $r_{ij} = \pi_{ij}/\pi_i^0$. Models explicitly devised for the π_{ij} are also implicit models for the r_{ij}, and vice versa, a fact to which we shall often refer. The categories of O and D are deliberately referred to as statuses, that is, positions in social structure broadly defined, although clearly in some cases this designation may not be appropriate.

Because the origin status variable O is logically prior to the destination status variable D, it is natural to attempt to characterize the variable that intervenes between O and D. This intervening variable may be denoted as X, and we hasten to point out that X might actually be composed of several variables, each of which might be important in its own right as a (partial) intervening variable. For the present argument, X will refer to the composite representation of all intervening variables, a single variable formed in an appropriate way by considering the joint variable with components equal to each of the intervening variables. Standard methods imply that, if X can be adequately specified and properly measured, the association between O and D will disappear when X is controlled. This means that the association between O and D would be nil in each conditional table formed by taking the cross-classification of O and D at each level of X.

If X were observable and discrete, standard contingency-table methods could be used to test whether X has been properly specified, an elementary log-linear model (O by $D \mid X$) would be of interest. If O, D, and X were quantitative and X were observable, standard methods appropriate for linear models could be used instead. This type of analysis would focus on the partial regression coefficient $\beta_{DO|X}$ and would lead to the conclusion that X is properly specified whenever $\beta_{DO|X} = 0$. Yet another approach to specifying the X variable that intervenes between O and D is the latent structure approach on which this chapter is based. Variables O, D, and X will be regarded as discrete (note that O and D are always discrete in the mobility-table context), but X will be conceived as latent, or inherently unobservable. The latent structure approach is distinguished from a linear model approach only by (1) positing discreteness for all variables and (2) positing that the intervening variable X is latent or incapable of being infallibly measured in a direct way. The latent structure approach allows a characterization of the intervening variable X without bringing additional information into the analysis, and characterizing this variable can be viewed as the primary objective of this research.

Let us suppose that X has T categories; the categories of X will be indexed by t, with t ranging from 1 to T, and these categories are referred to as the classes of X. The use of the term *class* is deliberate, just as the term *status* was deliberately chosen to refer to the categories of O and D. It will be seen later that the classes of X are, in general, mixtures of the statuses of O and D. Certain sets of statuses will serve to define these classes, though, strictly speaking, only in probabilistic terms. A major objective of the analysis will be to determine precisely which sets of statuses serve to define the respective classes of X. What is being suggested is that the pattern of mobility observed in the $O \times D$ mobility table can serve to define classes, or groups of statuses, within which the structure of mobility takes on a special character. In most respects, this approach will define a class of persons as a group that possesses random mobility chances with respect to the statuses that together constitute the particular class. That is, the absence of structured barriers to mobility among statuses within a class is the defining characteristic of a latent class of individuals. The close isomorphism between this statistical definition of a latent class and at least some sociological concepts of social class is perhaps the principal justification for the approach presented herein.

Let π_{ijt}^{ODX} refer to the expected proportion in the (i, j, t) cell of the indirectly observed O by D by X cross-classification. We assume that

$$\pi_{ij} = \sum_{t=1}^{T} \pi_{ijt}^{ODX} \tag{4.1}$$

which expresses the idea that the observable proportions p_{ij} are obtained by collapsing, or taking marginals over, the latent classes of X. If X is the variable intervening between origin status (O) and destination status (D), then the following relationship will also hold:

$$\pi_{ijt}^{ODX} = \pi_t^X \pi_{it}^{\bar{O}X} \pi_{jt}^{\bar{D}X} \tag{4.2}$$

for all i, j, and t. In Eq. (4.2), π_i^X refers to the probability that a randomly chosen member of the population will be a member of the t-th latent class, that is, the proportion of the population in latent class t. The parameter $\pi_{it}^{\bar{O}X}$ refers to the conditional probability that a member of the t-th latent class will have origin status i (for $i = 1, \ldots , I$), and the parameter $\pi_{jt}^{\bar{D}X}$ refers to the conditional probability that a member of the t-th latent class will have destination status j (for $j = 1, \ldots , I$). Equation (4.2) states that O and D are conditionally independent, given the levels of X, as should be the case if X intervenes between O and D. Note should be taken of the meaning of the parameters on the right-hand side of Eq. (4.2) in the mobility-table context: The π_t^X provide the distribution of the classes indirectly observed in the mobility table, and the $\pi_{it}^{\bar{O}X}$, $\pi_{jt}^{\bar{D}X}$ describe the distribution of the statuses within each class. Substituting Eq. (4.2) into Eq. (4.1) produces the fundamental equation of latent structure analysis:

$$\pi_{ij} = \sum_{t=1}^{T} \pi_t^X \pi_{it}^{\bar{O}X} \pi_{jt}^{\bar{D}X} \tag{4.3}$$

(Lazarsfeld & Henry, 1968; Goodman, 1974a).

Some Deductions from the Model

Let $\pi_{ijt}^{OD\bar{X}}$ denote the conditional probability that a person who has the (i,j) mobility pattern will be a member of the t-th latent class. Applying the definition of conditional probability gives

$$\pi_{ijt}^{OD\bar{X}} = \frac{\pi_{ijt}^{ODX}}{\pi_{ij}} \tag{4.4}$$

In general, $\pi_{ijt}^{OD\bar{X}}$ could be greater than zero for more than one value of t, implying that persons with a given mobility pattern (i,j) might be drawn from more than one latent class. A rule must be devised that provides an "optimal" prediction of latent class membership for any particular mobility pattern. Let t refer to the class of X for which $\pi_{ijt}^{OD\bar{X}}$ is at its maximum (i.e., t is the modal class of X for the $[i,j]$ mobility pattern). One prediction rule suited for the problem at hand is to assign all individuals with the (i,j) mobility pattern to the t-th class. When this is done, the expected proportion misallocated for the (ij) mobility pattern is

$$E_{ij} = 1 - \pi_{ijt}^{OD\bar{X}} \tag{4.5}$$

and the expected proportion misallocated into the latent classes over the whole table would be

$$E_2 = \sum_{ij} E_{ij} \, \pi_{ij} \tag{4.6}$$

The proportion correctly allocated into the latent classes given a model for the π_{ij} is merely $1 - E_2$. If the distribution of X were known in advance, an unconditional assignment of individuals into the classes of X could also be made. If t^* were the class of X where π_t^X is at a maximum, an assignment rule similar to that discussed above would place all individuals in the t^*-th latent class, irrespective of their mobility patterns. The expected proportion of errors with this assignment would be

$$E_1 = 1 - \pi_{t^*}^X \tag{4.7}$$

A measure of association seems appropriate as an index of the degree to which errors in predicting class membership are reduced when mobility patterns are taken into account. A proportional-reduction-in-error measure consistent with the definition of E_1 and E_2 above is the asymmetric λ,

$$\lambda_{X|OD} = \frac{(E_1 - E_2)}{E_1} \tag{4.8}$$

(see Goodman & Kruskal, 1954). As suggested by Clogg (1979), these measures can be of use in assessing the quality of a latent structure model in terms different from usual criteria based on goodness of fit of the model to data.

Returning now to the model parameters in Eq. (4.2), we note that π_{it}^{OX} appears to be defined in a way inconsistent with the specification of X as an intervening variable consequent to the origin status variable O, because it defines the conditional probability that the antecedent variable O takes on status i given that the consequent variable X is at level t (i.e., class t). We must quickly point out that X might actually be conceived as coexistent in time with O and one or more of which are consequent to O. Whatever the interpretation given to X (i.e., as coexistent in time with O, partly coexistent in time with O, or entirely consequent to O), the important idea is that the π_{it}^{OX} cannot be interpreted as simple "recruitment" probabilities, as is customary with the general latent structure model in many substantive contexts (see Lazarsfeld & Henry, 1968).

Let π_{it}^{OX} denote the conditional probability that a member of the i-th origin status will be in the t-th latent class. Clearly, this can be written as

$$\pi_{it}^{\tilde{O}X} = \frac{\pi_{it}^{OX}}{\pi_i^{O}} = \frac{\pi_{it}^{OX}\pi_t^{X}}{\pi_i^{O}} \tag{4.9}$$

Now the mobility rates $\pi_{ij} = \pi_{ij}/\pi_i^{O}$ can be written as

$$r_{ij} = \sum_{t=1}^{T} \pi_{it}^{\tilde{O}X}\pi_{jt}^{\tilde{D}X} \tag{4.10}$$

using Eqs. (4.9) and (4.3). Equation (4.10) says that the probability of moving from origin status i to destination status j is the sum of the T probabilities of moving from origin status i to latent class t to destination status j, where the moves (status i to class t) and (class t to status j) are independent of each other, for $t = 1, \ldots, T$. Equation (4.10) is similar to the Chapman–Kolmogorov identity for Markov chains (Feller, 1968, p.383), and it is clear that Eqs. (4.10) and (4.3) imply each other. The $\pi_{it}^{\tilde{O}X}$ can be interpreted as recruitment probabilities (i.e., they are the probabilities that a member of origin status i is recruited into the t-th latent class), and the $\pi_{jt}^{\tilde{D}X}$ can also be so interpreted (i.e., they are the probabilities that a member of class t will be recruited into the j-th destination status). The latent structure model thus bears some resemblance to a Markov-type model, with the important specification that the intervening states in the transition from status i to status j are unobservable or latent. The observable mobility rates r_{ij} are mixtures of the transition rates $\pi_{it}^{\tilde{O}X}$, $\pi_{jt}^{\tilde{D}X}$ derived from the latent structure model.

Some Special Cases

Let us now consider some special cases of the latent structure model defined in Eqs. (4.1)–(4.3). First consider a model H_1 where $T = 1$, implying that $\pi_{it}^{\tilde{O}X} = \pi_i^{O}$, $\pi_{jt}^{\tilde{D}X} = \pi_j^{D}$. In this case, Eq. (4.3) can be written as

$$\pi_{ij} = \pi_i^{O}\pi_j^{D} \tag{4.11}$$

demonstrating that H_1 is the hypothesis of independence between O and D, usually described as a model of "perfect mobility." It could be said that H_1 posits the nonexistence of a latent class structure, because the "class variable" X is degenerate, possessing only one class. Mobility is completely random with H_1, depending only on the distribution of statuses at origin and destination, and it is precisely this feature that demands that the occupational statuses be interpreted as designating only one class. Random mobility patterns imply that no structured barriers to mobility exist, apart from those constraints implied by differences in the two relevant status distributions. When the number of classes T is greater than one, a class structure is posited to exist, and these cases are of most interest to us here. If $T = 2$, for example, Eq. (4.3) would imply that within each latent class mobility would be random, depending only on the marginal distribution of the statuses at origin and destination within each class (i.e., $\pi_{it}^{\bar{O}X}$, $\pi_{jt}^{\bar{D}X}$), and a "class barrier" between two classes would be said to exist.

Consider next a model where $T = I + 1$ and where $\pi_{it}^{\bar{O}X} = \pi_{jt}^{\bar{D}X} = 1$, for $i = 1, \ldots ,I$. Then Eq. (4.3) reduces to

$$
\pi_{ij} = \begin{cases} \pi_i^X + \pi_{I+1}^X \; \pi_{i,I+1}^{\bar{O}X} \; \pi_{j,I+1}^{\bar{D}X} & \text{for } i = j \\[2mm] \pi_{I+1}^X \; \pi_{i,I+1}^{\bar{O}X} \; \pi_{j,I+1}^{\bar{D}X} & \text{for } i \neq j \end{cases} \tag{4.12}
$$

using the fact that $\pi_{it}^{\bar{O}X} = \pi_{jt}^{\bar{D}X} = 0$ when I is not equal to t. The first I classes of X are actually deterministic status classes, because the π_i^X, for $i = 1, \ldots ,I$, define the proportions in particular latent classes who stay in status I with probability one. We can let

$$\alpha_i = \pi_{I+1}^X \; \pi_{i,I+1}^{\bar{O}X}$$

$$\beta_j = \pi_{j,I+1}^{\bar{D}X}$$

and

$$\gamma_i = 1 + \frac{\pi_i^X}{\alpha_i \beta_j}, \qquad \text{for } i = 1, \ldots ,I$$

and then Eq. (4.12) reduces to

$$
\pi_{ij} = \begin{cases} \alpha_i \beta_j \gamma_i & \text{for } i = j \\[2mm] \alpha_i \beta_j & \text{for } i \neq j \end{cases} \tag{3.13}
$$

When the γ_i are all nonnegative, the $(T + 1)$-class restricted latent structure is equivalent to the model of quasi-perfect mobility, and the parameters γ_i are the "new indices of immobility" presented by Goodman (1969,1972) to measure status inheritance.

Another model similar to the quasi-perfect mobility model occurs when $T = I + 2$ and when the same 0–1 restrictions on the conditional probabilities are imposed. In this case, Eq. (4.3) would imply

$$
\pi_{ij} = \begin{cases} \pi_i^X + \sum\limits_{t=I+1}^{I+2} \pi_i^X \, \pi_{it}^{\bar{O}X} \, \pi_{jt}^{\bar{D}X} & \text{for } i = j \\[3mm] \sum\limits_{t=I+1}^{I+2} \pi_i^X \, \pi_{it}^{\bar{O}X} \, \pi_{jt}^{\bar{D}X} & \text{for } i \neq j \end{cases} \tag{4.14}
$$

Such a model was introduced by Goodman (1974a) in a very different context
and was designated a quasi-latent structure. For the model described by Eq.
(4.14), there are I deterministic status classes, where status inheritance (or
immobility) occurs with probability one. However, two classes of latent movers are posited to exist (instead of just one class of latent movers), and that is
all that distinguishes this model from the quasi-perfect mobility model. When
considering the quasi-perfect mobility model as applied to various sets of
mobility, Pullum (1975) found it necessary to modify the model by allowing
for deterministic classes of persons having a characteristic form of mobility
(e.g., mobility from status 1 to status 2) to explain the clustering adjacent to
the main diagonal in the mobility table. The generalization of the quasi-perfect mobility model suggested in the quasi-latent structure of Eq. (4.14) is to
posit the existence of two probabilistic classes of latent movers. It will be seen
later that this generalization of the quasi-perfect mobility model fits the data
very well, providing an adequate accounting for the clustering that is typically
observed on or adjacent to the main diagonal.

To draw out more fully the relationships between the quasi-perfect mobility model and the quasi-latent structure model, let

$$\alpha_{i1} = \pi^X_{I+1} \, \pi^{OX}_{i,I+1}$$

$$\alpha_{i2} = \pi^X_{I+2} \, \pi^{OX}_{i,I+2}$$

$$\beta_{i1} = \pi^{\bar{D}X}_{j,I+1}$$

and

$$\beta_{i2} = \pi^{\bar{D}X}_{j,I+2}$$

With these quantities, Eq. (4.14) can be written as

$$\pi_{ij} = \begin{cases} \pi^X_i + \sum_{k=1}^{2} \alpha_{ik}\beta_{jk} & \text{for } i = j \\ \sum_{k=1}^{2} \alpha_{ik}\beta_{jk} & \text{for } i \neq j \end{cases} \tag{4.15}$$

letting $s_{ij} = \Sigma_k \alpha_{ik}\beta_{jk}$, we can write

$$\gamma_i = 1 + \frac{\pi^X_i}{s_{ij}} \tag{4.16}$$

and (14) and (15) can be rewritten as

$$\pi_{ij} = \begin{cases} s_{ij}\,\gamma_i^* & \text{for } i = j \\ s_{ij} & \text{for } i \neq j \end{cases} \tag{4.17}$$

Equations (4.15)–(4.17) show that γ_i^* is a measure of the status inheritance in status I (for $I = 1, \ldots, I$), relative to the quasi-latent structure model. That is, γ_i^* (which will be greater than or equal to 1, since π_i^X is greater than or equal to 0) measures the surplus of stayers in status I which cannot be accounted for by the expected immobility under a model positing two latent classes of movers.

5

Analyzing the Relationship between Annual Labor-Market Experiences and Labor-Force Positions:
A Modification of the Labor Utilization Framework

INTRODUCTION

In the last chapter we looked at latent class models of mobility in an attempt to understand the transitions from one occupation to another across generations. Here we develop a means for linking prior labor force activity to current labor force outcomes. We provide a major modification of the Labor Utilization Framework of Sullivan (1978), Clogg (1979), and Clogg and Sullivan (1983), which is well suited for the study of labor-force dynamics. This typology has two parts, one set of categories referring to labor-force behavior throughout the previous year, and one set of categories referring to currently held labor-force positions. By classifying the previous year's labor force activity as "origins" (or from where people might start a search for better positions), and current labor force position as "destinations," we can examine what the relationships are between where one starts and where one ends up in a portion of the stratification system defined by labor market activities and labor force outcomes. But instead of examining these transitions across generations (as we did in the previous chapter) we will examine changes across short spans of time among the same people.

We developed this approach to overcome a deficiency in most sociological studies of labor markets, namely the lack of attention paid to access to steady year-round, full-time employment. Most current research on the sociology of labor markets has generally not incorporated labor-force concepts or measures in the comprehensive manner suggested here. Indeed, working definitions of the scope of the field usually ignore or at least downplay the significance of labor-force concepts or measures based on them. The concepts and measures used in the classical works on the labor force (Durand, 1948; Hauser, 1949; Bancroft, 1958) or more recent works with similar orientations (Sullivan, 1978; Clogg, 1979; Schervish, 1983) are simply not given much attention.[1]

Given empirical and theoretical considerations thus far, it is apparent that it would be fruitful to refocus attention on labor-force behavior and on how such behavior is related to the distribution of labor-force positions. This will, in turn, enhance our understanding of the distribution of market wages, status, and other job rewards, and thus inequality in general. What appears to be needed is a framework that focuses on the dynamic aspects of the labor force, and one that recognizes the role of this dynamic in the distribution of labor-market rewards and inequality.

Measures of unemployment, part-time employment, subunemployment, or underemployment—or any other labor-force concepts—are conspicuously absent from existing sociological models of the process of stratification. In most cases where such measures are considered at all, they are viewed as selection criteria rather than labor-force outcomes that deserve to be explained (see Schervish, 1983, for an important exception to this generalization). Kalleberg and Berg (1987, p. xii) hint at this deficiency in their preface to an otherwise comprehensive synthesis of the field: "We have given insufficient attention to the *labor force* [their emphasis] and its human members." The framework we develop in this and subsequent chapters is designed to remedy this deficiency. We provide both substantive and statistical justification for this new approach, demonstrate how it can be used to uncover important trends, show that it leads to a picture of the U.S. labor market that emphasizes transience rather than persistence in key labor-force types, and offer concrete suggestions for using the approach in other research.

Labor-Market Experience Categories

Our first goal is to create a composite variable that summarizes the main types of job-search (or nonsearch) behavior. A scheme that culminated in 16 basic types was devised through exploratory methods in combination with substantive criteria. The final categories chosen to summarize labor-market experiences—labor-force event histories—were as consistent as possible with standard labor-force concepts.[2] Definitions of these types appear in Table 5.1.

Table 5.1
Categories of Prior Labor Market Experience

Category No.. and Name	Description (CPS Items and Codes Used)[a]
1. NW-NL	Nonworker, not looking (item 3 = 5; item 4 = 1)
2. NW-L15+	Nonworker, looked for 1–14 weeks (item 3 = 5; $4 \le$ item $4 \le 6$)
3. NW-L14	Nonworker, looked for work 1–14 weeks (item3 = 5; $2 \le$ item $4 \le 3$)
4. PTPY-NL	Part-time, part-year, not looking (item 3 = 4; item 5 = 1)
5. PTPY-L1(15+)	Part-time, part-year worker, looked 15+ weeks in one stretch (item 1 = 1; item 3 = 4; $5 \le$ item $5 \le 7$)
6. PTPY-L2(15+) ..	Part-time, part-year worker, looked 15+ weeks in 2 or more stretches ($2 \le$ item $1 \le 3$; item 3 = 4; $5 \le$ item $5 \le 7$)
7. PTPY-L1 (14) ...	Part-time, part-year worker, looked 1–14 weeks in one stretch (item 1 = 1; item 3 = 4; $2 \le$ item $5 \le 4$)
8. PTPY-L2(14)	Part-time, part-year worker, looked 1–14 weeks in two or more stretches ($2 \le$ item $1 \le 3$; item 3 = 4; $2 \le$ item $5 \le$ item 4)
9. FTPY-NL	Full-time part-year worker, not looking for work (item 3 = 3; item 5 = 1)
10. FTPY-L1 (15+)	Full-time, part-year worker, looked 15+ weeks in one stretch (item 1 = 1; item 3 = 3; $5 \le$ item $5 \le 7$)
11. FTPY-L2(15+)	Full-time, part-year worker, looked 15+ weeks in two or more stretches ($2 \le$ item $1 \le 3$; item 3 = 3; $5 \le$ item $5 \le 7$)
12. FTPY-L1(14)	Full-time, part-year worker, looked 1-14 weeks in one stretch (item 1 = 1; item 3 = 3; $2 \le$ item $5 \le 4$)
13. FTPY-L2(14)	Full-time, part-year worker, looked 1-14 weeks in two or more stretches ($2 \le$ item $1 \le 3$; item 3 = 3; $2 \le$ item $5 < 4$).
14. PTFY-OTHER ..	Part-time, full-year worker, voluntary (item 3 = 2; item 2 not equal to 1 or 3).
15. PTFY-INVOL ...	Part-time, full-year worker, involuntary (item 3 = 2; item 2 = 1; or item 2 = 3).
16. FTFY	Full-time, full-year worker (item 3 = 1)

Note: Categories 1, 2-15, and 16 correspond to the stable inactive, unstable active, and stable active categories respectively.
[a]See Appendix for CPS items and codes used. Sorting is hierarchical as indicated.
Taken from Clifford C. Clogg, Scott R. Eliason, & Robert J. Wahl, 1990, "Labor Market Experiences and Labor Force Outcomes." *American Journal of Sociology,* p. 1540.

The two extremes in labor-force behavior are represented in the first and last categories of the experience typology. Category 1 (*nonworker, not looking*) can be called the stable inactive category of the population. People in this category did not work or look for work in the previous year. This is analogous to the concept of "economically inactive" people (Bancroft, 1958). In most official labor-force statistics as well as in sociological and demographic studies the economically inactive are defined with reference to a particular time, such as the week of the given labor-force survey, rather than to an interval of time as long as a year.[3]

Category 16 (full time, full year) can be called the *stable full-time active* category. Members in this category worked full time for the entire year. Some may have shifted from one full-time job to another without any recorded interruptions in employment history. It is important to note that this category does *not* represent job stability or occupational stability, but rather *stability in employment*. This category contains some people who searched for alternative employment or who were promoted to or demoted from jobs held at the beginning of the year, but this search behavior or job mobility did not include periods of unemployment or part-time work. Most models of earnings and occupational attainment are developed for people in this category.

Categories 2–15 represent intermediate levels of stability or instability. As a whole, this set of workers are "reserve" labor, representing marginal workers of varying types. The search behavior, unemployment spells, and marginal worker status associated with these categories reflect labor-market forces that are important to take into account when we study labor-force dynamics. It is important to realize that these intermediate categories are not necessarily ordered from low to high. Instead, the experience levels should be viewed as a partially ordered categorical variable; the extremes appear to be well established, but the intermediate levels cannot be ordered in advance.

The 16 types provide a comprehensive but complex picture of labor-force behavior throughout the previous year. Nevertheless, these 16 types can be simplified and collapsed into three natural categories, which we call the *stable inactive* (level 1), the *unstable active* (levels 2–15), and the *stable full-time active* (level 16). In each case, activity or stability is defined with reference to behavior throughout a calendar year.

Labor-Force Positions (Current Status)

Labor-force positions are measured using a modified version of the Labor Utilization Framework (LUF) proposed in Clogg, Sullivan, and Mutchler (1986). Recall that the LUF scheme measures underemployment according to criteria that pertain to currently held positions. The modified method used here disaggregates unemployment and part-time employment into different types while maintaining the other original LUF categories (see earlier chapters, as well as Sullivan, 1978, and Clogg, 1979).

The labor-force position variable has 10 categories that refer to current labor-force status:

1. Not in the labor force
2. Unemployed—new entrants and reentrants (unemployment not accounted for by the next two types of unemployment)
3. Unemployed—quits, job losses
4.. Unemployed—layoffs
5. Part-time employed (low hours)—involuntary (those for whom no full-time work is available)

6. Part-time employed (low hours)—economic reasons (e.g., slack work)
7. Part-time employed (low hours)—voluntary
8. Underemployed by low income (earnings)—see earlier chapters or Clogg & Sullivan, 1983
9. Mismatch—overeducated for current occupation (see earlier chapters or Clogg & Shockey, (1984b)
10. Adequate full-time—full-time workers with adequate income not counted above

All standard labor-force measures of current status can be derived from these categories, in some instances by combining categories (Clogg et al., 1986). As we have described above, the low-income and mismatch categories are unique to LUF. Finally, recall that the low-income category is actually based on a comparison of the average weekly wage of the previous year (adjusted for weeks worked) with the poverty thresholds developed by the Social Security Administration.[4]

Because of the temporal ordering of prior experience and current position, the two variables can be cross-classified to form a transition matrix. This matrix provides a link between past labor-force behavior and current labor-force position rather than the more customary link between positions at two points in time. Our basic claim is that this contingency table can serve as a definition of the labor-market matching process. The image of the labor-market matching process that this table provides stands in sharp contrast to the occupation or occupation-by-industry matching processes implicit in most sociological models, which usually ignore labor-force behavior.[5]

LABOR-MARKET EXPERIENCES AND LABOR-FORCE OUTCOMES IN A RECESSION

In this section we analyze the association between experience types measured for calendar year 1981 and labor-force outcomes in March 1982. This time interval was selected because the deepest recession since the 1930s was under way in 1981. Recession shocks produced high rates of instability, which is reflected in the distribution of people across experience categories. Recession effects are also apparent in the exceptionally high rates of unemployment and underemployment in 1982. Our initial assumption was that this interval would be the best possible case for observing the labor-force transitions that accompany labor-market processes adjusting to economic stress. The results pertain to the entire civilian, noninstitutionalized population age 16 and over, so practically everyone "exposed" to recession shocks is represented in the sample used. The 16 × 10 cross-classification of experience by position appears in Table 5.2.[6]

Table 5.2
Cross-Classified Prior Labor Market Experiences and Current Labor Force Outcomes, March 1982 (Weighted Frequencies)

Labor-Market Experience Category[a]	Labor-Force Category (March 1982)[b]									
	NILF	UR	UQ	UL	HI	HE	HV	I	M	ADFT
1. NW-NL	35,113	689	172	27	97	50	530	0[c]	84	455
2. NW-L15+	269	255	259	27	40	5	22	0[c]	18	81
3. NW-L14	741	186	28	8	8	6	36	0[c]	14	47
4. PTPY-NL	3,442	265	107	43	295	182	2,775	890	113	400
5. PTPY-L1(15+) . . .	148	70	131	35	88	63	172	131	15	59
6. PTPY-L2(15+) . . .	108	94	133	34	59	86	81	75	10	69
7. PTPY-L1 (14)	277	60	63	37	102	61	388	200	25	92
8. PTPY-L2(14)	116	55	48	17	51	46	174	91	21	61
9. FTPY-NL	2,750	205	236	95	79	199	518	578	670	2,631
10. FTPY-L1(15+)	232	112	535	192	82	99	61	192	188	827
11. FTPY-L2(15+)	168	93	486	211	68	111	53	144	106	464
12. FTPY-L1(14)	350	90	470	302	73	175	162	320	434	1,734
13. FTPY-L2(14)	119	52	186	155	44	113	43	132	144	677
14. PTFY-OTHER . . .	418	39	73	18	122	99	3,116	782	178	517
15. PTFY-INVOL	47	15	36	19	303	235	102	227	44	146
16. PTFY	714	100	564	426	152	932	1,019	2,322	8,980	31,512

[a]See Table 5.1 and Appendix 5.1 for derivation of categories of labor-market experience.
[b]See Clogg, Sullivan, and Mutchler (1986) and Clogg and Sullivan (1983). for definitions of underemployment categories used here.
[c]Structural zeros: persons in the first three categories of the EXPER variable do not have income to report for the previous year.
Taken from Clifford C. Clogg, Scott R. Eliason, & Robert J. Wahl, 1990, "Labor Market Experiences and Labor Force Outcomes." *American Journal of Sociology*, p. 1545.

Outflow Rates

Table 5.3 gives the outflow percentages, the conditional probabilities of being in current labor-force positions for each level of prior experience. These percentages would be highly similar as one looks across each row is there was no relationship between prior experience and current labor force position. Obviously, there is some relationship between prior labor force experience and current labor force position, and this can be seen by looking at rows 1 and 16 in Table 5.3.

Row 1 contains those who were not working and not looking for work in 1981. Almost 95% of this category are not in the labor force in 1982 either. Most of the remainder experience some type of part-time or underemployed status in the survey week in March. Only 1.2% of this category move from economic inactivity to an adequate full-time labor-force status (column 10 of Table 5.3). Next, consider row 16 (full-time, full-year workers). This group's essential stability is indicated by the fact that about two-thirds of its members (67.4%) are currently located in the adequate, fulltime employment category,

Table 5.3

Outflow Percentages From Prior Experience to Current Labor Force Positions Based on the Frequencies in Table 5.2

Labor-Market Experience Category	Percentage inLabor-Force Category									
	NILF	UR	UQ	UL	HI	HE	HV	I	M	ADFT
1. NW-NL	94.4	1.9	.5	.1	.3	.1	1.4	.0	.2	1.2
2. NW-L15+	27.5	26.1	26.5	2.8	4.1	.6	2.2	.0	1.9	8.3
3. NW-L14	69.0	17.3	2.6	.8	.7	.6	3.3	.0	1.3	4.4
4. PTPY-NL	40.4	3.1	1.3	.5	3.5	2.1	32.6	10.5	1.3	4.7
5. PTPY-L1(15+) . .	16.2	7.7	14.3	3.8	9.6	6.9	18.8	14.4	1.7	6.5
6. PTPY-L2(15+) . .	14.5	12.5	17.7	4.5	7.9	11.5	10.8	10.1	1.3	9.2
7. PTPY-L1 (14) . . .	21.2	4.6	4.8	2.8	7.8	4.7	29.7	15.3	1.9	7.1
8. PTPY-L2(14) . . .	17.0	8.1	7.1	2.6	7.5	6.8	25.6	13.3	3.0	9.0
9. FTPY-NL	34.5	2.6	3.0	1.2	1.0	2.5	6.5	7.3	8.4	33.0
10. FTPY-L1 (15+) . . .	9.2	4.4	21.2	7.6	3.2	3.9	2.4	7.6	7.5	32.8
11. FTPY-L2(15+)	8.8	4.9	25.6	11.1	3.6	5.8	2.8	7.5	5.6	24.4
12. FTPY-L1(14)	8.5	2.2	11.4	7.3	1.8	4.3	3.9	7.8	10.6	42.2
13. FTPY-L2(14)	7.2	3.1	11.2	9.3	2.6	6.8	2.6	8.0	8.6	40.7
14. PTFY-OTHER . . .	7.8	.7	1.4	.3	2.3	1.8	58.1	14.6	3.3	9.6
15. PTFY-INVOL	4.0	1.3	3.1	1.6	25.8	20.1	8.7	19.3	3.7	12.4
16. FTFY	1.5	.2	1.2	.9	.3	2.0	2.2	5.0	19.2	67.2

Note: Percentages are calculated from weighted frequencies in table 2. Row totals may not equal 100.0 owing to rounding off. Fractional frequencies rather than rounded frequencies were used to calculate these percentages.
Taken from Clifford C. Clogg, Scott R. Eliason, & Robert J. Wahl, 1990, "Labor Market Experiences and Labor Force Outcomes." *American Journal of Sociology,* p. 1545.

with another 19.2% in the mismatch (overeducation) status. In other words, nearly 90% of those in this prior experience category persist in a full-time status and about 7% were downgraded by movement to current positions that represent underemployment, which highlights the effects of the recession in the early 1980s.

The outflow rates for the middle 14 categories of prior experience are closer to independence, but even so there are still substantial differences among types. For example, nearly 60% of the part-time full-year category "move" to the voluntary part-time category current position (column 7 of Table 5.3). Types 2–8 and 14 have very high rates of underemployment but also account for less than 10% of the adequate full-time category. Categories 9–13 have high rates of underemployment or part-time work but also moderate percentages of adequately employed full-time workers.

Outflow rates like those in Table 5.3 can be used to characterize the labor-market process that matches prior labor-force histories to current labor-force positions. However, these rates are very heterogeneous across experience levels, which argues for some kind of modeling strategy that captures salient aspects of this variability in terms of a few parameters.

Inflow Rates

The inflow rates for each current position category are presented in Table 5.4. These quantities describe how each current labor-force position recruits members from the set of possible experience categories. Considering the inflow rates on just the two extremes of the position measure shows a much higher degree of transience than would probably be expected. The extremes on the current position measure (not in the labor force and adequate fulltime employment) are the easiest to characterize. Nearly 80% of the March 1982 not-in-the-labor-force category were in the not-working/not-in-the-labor-force category in the prior year. But this implies that a full 20% of those currently not in the labor force have had some measurable labor-force activity during the previous year, which gives some indication of the transient character of the not-in-the-labor-force category often used in contemporary research. Nearly 80% of those currently in the adequate full-time employment category are drawn from the full-time, full-year worker prior experience category. Stated another way, about one-fifth of those currently in this category are drawn from experience types that reflect some type of marginal-worker status over the previous year, which indicates the transient character of this status. In

Table 5.4

Inflow Percentages from Prior Experience to Current Labor Force Positions Based on the Frequencies in Table 5.2

Labor-Market Experience Category	Percentage in Labor-Market-Experience Category within Each Labor-Force Category									
	NILF	UR	UQ	UL	HI	HE	HV	I	M	ADFT
1. NW-NL	78.0	28.9	4.9	1.6	5.8	2.0	5.7	.0	.8	1.1
2. NW-L15+	.6	10.7	7.3	1.7	2.4	.2	.2	.0	.2	.2
3. NW-L14	1.6	7.8	.8	.5	.5	.2	.4	.0	.1	.1
4. PTPY-NL	7.6	11.1	3.0	2.6	17.8	7.4	30.0	14.6	1.0	1.0
5. PTPY-L1(15+)	.3	2.9	3.7	2.1	5.3	2.6	1.9	2.2	.1	.1
6. PTPY-L2(15+)	.2	3.9	3.8	2.0	3.6	3.5	.9	1.2	.1	.2
7. PTPY-L1 (14)	.6	2.5	1.8	2.2	6.1	2.5	4.2	3.3	.2	.2
8. PTPY-L2(14)	.3	2.3	1.4	1.1	3.1	1.9	1.9	1.5	.2	.2
9. FTPY-NL	6.1	8.6	6.7	5.8	4.7	8.1	5.6	9.5	6.1	6.6
10. FTPY-L1 (15+)	.5	4.7	15.2	11.7	4.9	4.0	.7	3.2	1.7	2.1
11. FTPY-L2(15+)	.4	3.9	13.8	12.8	4.1	4.5	.6	2.4	1.0	1.2
12. FTPY-L1(14)	.8	3.8	13.3	18.3	4.4	7.1	1.8	5.3	3.9	4.4
13. FTPY-L2(14)	.3	2.2	5.3	9.4	2.6	4.6	.5	2.2	1.3	1.7
14. PTFY-OTHER	.9	1.6	2.1	1.1	7.3	4.0	33.7	12.9	1.6	1.3
15. PTFY-INVOL	.1	.6	1.0	1.1	18.2	9.6	1.1	3.7	.4	.4
16. FTFY	1.5	4.2	16.0	25.9	9.2	37.9	11.0	38.2	81.3	79.2

Note: Column sums (rows 1–16) might not equal 100.0 owing to round-off error.
Taken from Clifford C. Clogg, Scott R. Eliason, & Robert J. Wahl, 1990, "Labor Market Experiences and Labor Force Outcomes." *American Journal of Sociology*, p. 1545.

short, the persistence of full-time, full-year work and total inactivity is much less than we often think.

The several types of unemployment, part-time employment, and other types of economic underemployment (Clogg 1979) are drawn from a quite heterogeneous assortment of prior labor-force experiences. The marginal-worker types located between the not-in-the-labor-force and adequate fulltime, full-year employment are generally linked to the unstable active (intermediate) categories prior experience. But transience rather than persistence appears to be the norm for workers with marginal status.

If we simply collapse categories, nearly one-third of the population of working ages is located in some kind of marginal-worker status in terms of either recent behavior or current position, while another one-third is located in a status that is best described as economically inactive. It should be obvious that this framework is not forcing us to consider labor-force types that are unimportant. Even if we focus exclusively on the economically active, our results suggest that over 40% of this total should be regarded as marginal workers.[7]

In sum, the prior experience by current position contingency table contains new information about the labor-market matching process. This is demonstrated by (1) marginal distributions showing that marginal labor force participants are relatively abundant and should not be ignored, (2) outflow and inflow rates, showing varying levels of persistence and transience of types of economic activity, and (3) the strong association between prior experience and current position, which can be inferred from differentials in the outflow or inflow rates as well as from other quantities. Such information is simply unavailable when particular groups of workers or nonworkers are singled out for special attention and when other groups of workers or nonworkers are ignored. The evidence clearly indicates that prior experience and current position are not independent, implying that we have identified measurable sources of heterogeneity. On the other hand, the association is far from perfect; it is not the case that particular experience types map into one and only one position type. The relations between labor-market experiences and labor-force outcomes that can be inferred from a 16 × 10 contingency table are complex. In the rest of this section we try to reduce the complexity through special modeling tools well suited for such data.

A MODEL FOR ASSOCIATION

Standard mobility table modelling techniques will not work in summarizing the relationship between prior labor force experience and current position because the categories of each measure are not comparable. If we had data at multiple points in time (say, for the entire decade of the 1970s or 1980s) we could look at the relationship between labor force experience and labor force experience in prior years. But this would not tell us how prior experience is related to current labor force position. We could also look at the relationship between current position and current position measured in prior years, but this would not include information on labor force behavior *throughout* the year. In effect we would be comparing two snapshots 1 year apart. The only way to relate the effects of prior labor force behavior on where one is presently in the labor force is by examining the relationship between prior experience and current labor force position.[8]

THE MATCHING PROCESS SUMMARIZED BY ASSOCIATION MODELS

Goodness of Fit and Strength of Association

Table 5.5 gives likelihood-ratio chi-square statistics from several models applied to the 16 × 10 prior experience × current position table and to a 16 × 9 table obtained by deleting those currently not in the labor force. (See Appendix 5.2 for details on the models used in this analysis.) We want to compare the results with those outside the labor force not included to see what the consequences are of including only current labor participants in the analysis.

Our basic conclusion from Table 5.5 is that the association between experience types and labor-force positions must be characterized by at least two latent dimensions. The one-dimensional RC(1) model is simply not congru-

Table 5.5
Chi-Square Values from Association Models Applied to Data in Table 5.2

	16 × 10 Table (Total Population)				16 × 9 Table (Deleting NILF Category)		
	L^2	df	Percentage of Base[b]	D	L^2	df^a	Percentage of Base[b]
RC(0)[c]	146,522	132	—	45.4	49,869	117	—
RC(1)[d]	21,180	109	85.5	10.0	10,738	95	78.5
RC(2)[e]	8,728	88	94.0	5.1	2,220	75	95.5
RC(3)[f]	728	69	99.5	1.1	364	57	99.3

[a]Degrees of freedom, taking account of three structural zeros (as indicated in Table 5.1).
[b]Percentage of L^2 (RC[0]) accounted for by model.
[c]Independence model or null association model.
[d]RC(1) association model: one "dimension" of row-column association.
[e]RC(2) association model: two dimensions of row-column association.
[f]RC(3) association model: three dimensions of row-column assocation.
Taken from Clifford C. Clogg, Scott R. Eliason, & Robert J. Wahl, 1990, "Labor Market Experiences and Labor Force Outcomes." *American Journal of Sociology*, p. 1551.

Table 5.6
Association between Prior Experience and Current Position in
1982, Parameters from the RC(2) Association Model

	First Dimension	Second Dimension
Scores for types of Labor-market Experiences (values from i)		
1..	−1.28	−.61
2..	−.25	.94
3..	−.74	.26
4..	−.66	1.81
5..	− .09	1.78
6..	−.01	1.52
7..	−.24	1.88
8..	−.12	1.75
9..	−.03	−.33
10......................................	.35	.67
11......................................	.32	1.00
12......................................	.46	.40
13......................................	.51	.61
14......................................	−.12	2.67
15......................................	1.39	3.54
16......................................	1.07	−.48
Scores for types of Labor-force Outcomes (values of j);		
1..	−1.22	−.45
2..	−.51	.70
3..	.27	1.15
4..	.52	1.13
5..	.36	2.59
6..	.72	1.63
7..	−.21	2.56
8..	.37	1.40
9..	1.08	−.49
10......................................	.99	−.57
Intrinsic Association[a]		
ϕ......................................	1.583	.404
ρ......................................	.856	.570

[a]See text for definition of ϕ parameters. The ρ values are correlations between row and column scores in the first and second dimensions, respectively.
Taken from Clifford C. Clogg, Scott R. Eliason, & Robert J. Wahl, 1990, "Labor Market Experiences and Labor Force Outcomes." *American Journal of Sociology*, p. 1553.

ent with the data. Results from the 16 × 9 table lead to similar inferences about the number of dimensions required for capturing the association.

Parameter values for the RC(2) model appear in Table 5.6. The association between prior experience and current position is substantial. The intrinsic association in the first dimension is $\phi_1 = 1.583$. This is an estimate of a log-odds ratio, and it implies an odds ratio of 4.87 and a Yule's Q of 0.66. The Pearson correlation between row scores and column scores in the first dimension is $\rho_1 = 0.856$, which shows the substantial association in another metric. The association in the second dimension is also substantial: $\phi_2 = 0.404$ and $\rho_2 = 0.570$.

Interpretation of Score Parameters

Consider the first-dimension scores for the categories of prior experience, points along a continuum of labor-market experience as calibrated with information on labor-force outcomes. Earlier we noted how the two extreme categories of prior experience should denote extreme points in a scale derived from the 16 current position categories. This is almost borne out by the results, with $\mu_{1,1} = -1.28$ (not working–not looking) and $\mu_{16,1} = 1.07$ (fulltime, full-year work). The exception is that $\mu_{15,1} = 1.39$, which is slightly greater than the score for the last prior experience category. The 15th prior experience category (part-time, full-year workers who were parttime for involuntary reasons) is thus "out of order" with respect to the first dimension scores; however, this does not appreciably affect inferences reported below because this type represents less than 1% of the sample.

These scores appear to reflect distinctions among types of labor-force participation or types of instability in employment when they are regrouped as follows

Group 1 = no work, not seeking a job—EXPER level 1 (score = –1.28)
Group 2 = no work, seeking a job—levels 2, 3 (scores = –0.25, –0.75)
Group 3 = part-time and part-year work, not seeking a job—level 4
      (score = –0.66)
Group 4 = part-time and part-year work, seeking a job—levels 5–8
      (scores = –0.09, –0.01, –0.24, –0.12)
Group 5 = part-time or part-year work, voluntary or not seeking a job—
      levels 9, 14 (scores = –0.12, –0.03)
Group 6 = part-time or part-year work, involuntary or seeking a job—
      levels 10–13 and 15 (scores = 0.35, 0.32, 0.46, 0.51, 1.39)
Group 7 = full-time, full-year work—level 16 (score = 1.07)

We conclude that the first dimension of prior experience scores indicate a reasonably well constructed a priori ordering of the experience variable, with the possible exception of the score for the 15th category. They provide a natural clustering of the types of labor-force behavior recognized in the composite measure of prior experiences.

Scores for the second dimension can also be interpreted in a relatively straightforward manner. Divide the prior experience categories along two dimensions: full-year versus part-year worker and full-time versus part-time employed. When this is done, the second-dimension scores represent contrasts among these four basic types of workers.[9] The second-dimension scores thus represent contrasts among levels of two of the specific indicators used to define prior labor force experience (see the Appendix), which gives further meaning to the second dimension.

The scores for the current position measure are also interpretable in terms of the matching process conceived here. The not-in-the-laborforce category (position = 1) has score $v_{1,1} = -1.22$, which is the smallest of the 10 scores. The mismatch and adequate, full-time categories have associated scores of 1.08 and 0.99, respectively, which are the largest scores. The higher score for the former category is partly a consequence of the fact that occupational mismatch can be present only for workers with at least some post-high school education. This implies a rather favorable labor-force position for mismatched workers, at least when calibrated against the types of instability and marginality built into our typology. In fact, the mismatch category could be combined with the adequate employment category becuase their scores are nearly equal on both dimensions.[10] Both the ordering and the relative spacing of current position categories in the first dimension seem plausible, and we regard this as evidence for the basic validity of the model and the contingency table to which it has been applied.

The second-dimension scores represent a contrast between economic underemployment (current position levels 2–8, mean score = 1.59) and full-time employment (current position levels 9 and 10, mean score = –0.53). Both dimensions of the current position measure tap distinct aspects of current labor force activity; the first dimension places the different categories of our measure on a continuum anchored at each end by those not at work and not seeking a job and those in full-time/year-round work, and the second dimension highlights the contrast between full-time employment and economic underemployment.

Geometric Representation of the Contingency

A graphic summary of the score parameters and the interaction they depict appears in Fig. 5.1. It represents the geometry of the matching process and gives interpretations that are similar to those used in conventional clustering procedures. The first dimension scores are used to define the horizon-

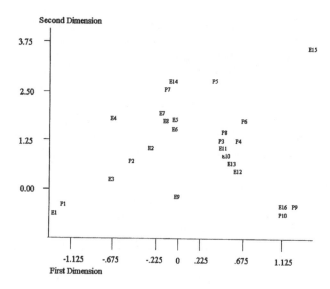

Figure 5.1. Plot of prior experience and current position scores from the RC(2) association model. P 1–10 refer to current position scores 1–10; E1–16 refer to prior experience scores 1–16. Take from Clifford C. Clogg, Scott R. Eliason, & Robert J. Wahl, 1990, "Labor Market Experiences and Labor Force Outcomes." American Journal of Sociology, p. 1556.

tal axis, and the second-dimension scores are used to define the vertical axis (for details see Goodman, 1986, and Eliason, 1995b).[11]

The graph shows that experience type 15 is an extreme outlier; appearing by itself in the upper right-hand corner of the figure.[12] Experence type 1 (E1) and current position type 1 (P1) are almost inseparable. These two points appear by themselves in the lower left-hand corner. In other words, the stable-inactive category defined in reference to the previous year is virtually indistinguishable from the not-in-the-laborforce category defined in reference to the March measurement of current labor force position. It is also evident that current position category 10 (P10), representing adequate full-time employment, is located in virtually the same position in the space as E16 (= full-time, full-year worker) and P9 (= mismatch).

There are four clusters of points in the middle of the graph representing the close correspondence of experience types (E) and position types (P) within each cluster. These are

Group 1: E3, P2, and E2
Group 2: E4, E7, E8, E5, E6
Group 3: P7, E14, and P5
Group 4: E12, E13, E10, Ell, P3, P4, P8, P6[13]

In sum, we are encouraged by the ability of the model to (1) summarize the rather complicated association between prior experience and current labor force position, (2) score both prior experience and current position categories in an intuitively plausible manner, and (3) reveal, by graphic means and other ways, the very strong experience–position association present in the data. The model summarizes the labor-market matching process in a relatively simple fashion and several alternative ways to regroup categories have been suggested that can be used to simplify the task of using the information in this new contingency table in further analyses (see Appendix 5.4). We next begin assessing subgroup variability in the matching process in terms of this basic model.

FURTHER ANALYSES OF THE MATCHING PROCESS: PERSISTENCE, CHANGE, AND STRUCTURE

A Simple Trichotomy for Labor-Market Experiences

In this section, we present analyses based on a simple modification of the prior experience measure. One goal is to show how a simple trichotomized version of the variable performs as a summary measure. We believe that this admittedly crude measure can be justified to some extent as a simple strategy for summarizing labor-force behavior.

The trichotomy is obtained by combining categories 2–15 of the prior experience measure. The levels of the trichotomy correspond to *stable inactive* (level 1 of the 16 level experience variable), *unstable active* (levels 2 through 15 of the 16 level experience variable), and *stable full-time active* (level 16 of the 16 level experience variable). The category labels are used deliberately. Perhaps the middle category (unstable active) could be redefined in some way, but the essential property of this category is the labor force marginality of the group. The labor-market matching process has produced either unstable employment conditions or marginal worker status, perhaps temporary or "frictional," for all members of this category.

Inflow and Outflow Rates Based on the Trichotomous Labor Force Experience Measure

Table 5.7 presents the (weighted) cell frequencies, outflow rates (second entry in each cell), and inflow rates (third entry in each cell) for the trichotomous labor force experience measure cross-classified by current labor force position. The marginal distribution of our new experience measure splits the population of working ages into approximately equal parts, 31% stable inac-

Table 5.7
Simplified Labor Market Experience by Current Position, March 1982 (Weighted Frequencies)

Experience Category	Labor-Force Category									Adequate Full-Time	Total
	NILF	UR	UQ	UL	HI	HE	HV	I	M		
Stable inactive:											
Weighted frequency	35,821	689	172	27	97	50	530	0	706	455	37,839
Row percentage	92.8	1.8	.5	.1	.3	.1	1.4	0	1.9	1.2	30.6
Column percentage	78.0	28.9	4.9	1.6	5.8	2.0	5.7	0	6.1	1.1	
Unstable inactive:											
Weighted frequency	9,185	1,591	2,791	1,193	1,412	1,480	7,703	3,763	1,980	7,805	38,902
Row percentage	23.6	4.1	7.2	3.1	3.6	3.8	19.8	9.7	5.1	20.1	31.5
Column percentage	20.4	66.9	79.1	72.5	85.0	60.1	83.3	61.8	17.0	19.6	
Stable full-time:											
Weighted frequency	714	100	564	426	152	932	1,019	2,322	8,980	31,512	46,721
Row Percentage	1.5	.2	1.2	.9	.3	2.0	2.2	5.0	19.2	67.2	37.8
Column percentage	1.6	4.2	16.0	25.9	9.2	37.9	11.0	38.2	77.0	79.2	
Column total (%)	36.5	1.9	2.9	1.3	1.3	2.0	7.5	4.9	9.4	32.2	100.0

Note: Row percentages are outflow rates, and column percentages are inflow rates.
Taken from Clifford C. Clogg, Scott R. Eliason, & Robert J. Wahl, 1990, "Labor Market Experiences and Labor Force Outcomes." *American Journal of Sociology,* p. 1545.

tive, 32% unstable, and 38% stable full-time.[14] This has two important implications, one substantive and the other methodological. The substantive implication is that there is a remarkable degree of dynamism in the labor-market matching process. The high relative frequency in the unstable active category is an important sociological and demographic fact that has not been taken into account in most previous research. The methodological implication is that both the long and short version of the prior experience measure have numbers sufficient in each category to include in an analysis of the attainment process. The marginal distribution of both measures shows how severe the censoring or truncation actually is when one restricts attention to the stable (continuously employed) full-time labor force. Most formal models of earnings attainments select the sample from category 3 of our simplified labor force experience measure (stable full-time). Many models for occupational attainments select at least some of the persons in category 2, but then all sample members selected for modeling are essentially treated as though they derived from category 3. Respondents with no job at the time of the survey or those with no job information available have been excluded from status-attainment analyses (e.g., Jencks, Crouse, & Mueser, 1983; Hauser, Tsai, & Sewell, 1983). And for those respondents not currently employed at the time of the survey, the "most recent occupation" has been used to code values of occupational status (Duncan, Featherman, & Duncan, 1972). If we exclude from consideration the stable inactive category, then fully 45% of the "labor-force eligible" population will be censored out of models that focus only on the stable full-time category of workers.

Inflow and outflow rates for the stable inactive and for the stable fulltime categories have been discussed previously. The unstable active category is associated with a substantial rate of nonparticipation in March 1982 (23.6%), a very high rate of voluntary part-time employment (19.8%), and only a 20.1% rate of adequate full-time employment. Rates of underemployment are very high for this category, which is another indication of the marginal-worker status of this experience type. (Compare row percentages for the unstable active category with the marginal distribution of the current position categories.) Inflow rates indicate that about one-fifth (20.4%) of those not currently in the labor force are drawn from the unstable category and that about one-fifth (19.6%) of those currently adequately employed are drawn from this category. On the other hand, nearly two-thirds of each underemployment category (except mismatch) are drawn from the unstable active category of prior experience. A picture of dynamism in the labor-market matching process emerges, a picture that we believe has been blurred to a great extent by usual models that ignore labor-force dynamics.[15]

Stability and Change in Labor-Market Experiences over Time

We next examine the marginal distributions of our simplified experience measure for the total population and for males and females over time.[16] We look at the distribution across the three labor force experience categories from 1976 to 1984, a period marked by considerable economic restructuring (see Bluestone & Harrison, 1982). Table 5.8 summarizes the distribution among the three types of labor-market experiences, that is, stable inactive, unstable active, and stable full-time, first for the total population and then for males and females.

The most striking thing is the essential homogeneity of the marginal distributions over time.[17] For the total population, the proportion in the stable inactive category varies by only about 2% absolutely, from a low of 30.2% in 1979 to a high of 31.6% in 1976. The proportion unstable active varies to about the same degree, 30.1% in 1984 to 32.3% in 1976. The proportion stable full-time varies slightly more (about 3%), with the lowest figure in 1976 (36.1%) and the highest in 1984 (39.4%). Except for the recession effect on figures reported for 1981 and 1982, the unstable active category is remarkably stable over time for both males and females, with females registering about a 4% higher (absolute) figure than males. The chief exceptions to the above generalization are as follows: (1) for males there is a slight increase in the percentage stable inactive, from about 18.5% in the mid-1970s to about 21% in the

Table 5.8
Labor Market Experience Types for the Total Population
and By Gender, 1976–1984[a]

	Labor Market Experience Types (Trichotomy) (%)								
	Total			Males			Females		
Year	SI[b]	UA	SF	SI	UA	SF	SI	UA	SF
1976	31.6	32.3	36.1	18.6	30.2	51.3	43.4	34.2	22.9
1977	31.1	31.9	37.0	18.5	29.6	51.9	42.4	34.1	23.4
1978	30.2	31.1	38.5	18.5	28.1	53.4	41.3	33.7	24.9
1979	30.2	31.1	38.7	18.8	27.7	53.4	40.5	34.1	25.3
1980	30.5	31.4	38.1	19.4	28.8	51.7	40.5	33.7	25.8
1981	30.6	31.5	37.8	19.7	29.4	50.9	40.6	33.4	26.0
1982	31.1	32.2	36.7	20.4	31.3	48.3	40.8	33.1	26.1
1983	31.2	31.0	37.8	21.0	29.7	49.3	40.4	32.2	27.3
1984	30.5	30.1	39.4	21.0	27.5	51.5	39.1	32.4	28.4

[a]Distribution for year t obtained from the March CPS for year $t + 1$.
[b]SI = Stable inactive, UA = unstable active, SF = stable full-time, levels 1, 2, and 3, respectively of EXPER3.
Taken from Clifford C. Clogg, Scott R. Eliason, & Robert J. Wahl, 1990, "Labor Market Experiences and Labor Force Outcomes." *American Journal of Sociology*, p. 1545.

mid-1980s, but no major trend in the other categories, if we disregard the recession of 1981–82; and (2) for females there is a steady reduction in the percentage stable inactive, from about 43% to about 39%, and a steady and substantial increase in the percentage stable full-time.

The essential constancy in the unstable active category indicates that the aggregate prevalence of marginal-worker status, as measured by labor-force behavior over an entire year, can be regarded as a structural property of the U.S. labor market. For example, there appears to be little effect of recession or growth on the relative frequency of this category. We can also note that the unstable active proportion for females in the mid-1980s is approximately the same as for males in the mid-1970s, which gives an alternative way to gauge convergence between the sexes in labor-force behavior.

The above results are somewhat surprising because variability in current measures of underemployment and unemployment was much more dramatic over this interval. For example, the economic underemployment category, which is the sum of current position categories 2–8 and a proxy for discouraged workers, varied by about 8% over the same interval (Clogg & Shockey, 1985). Unemployment varied from a low of 6% (in 1979) to a high of over 10% (in 1982). That the variability in current measures of unemployment or underemployment is not matched at all by variability in the aggregate measures of stability/instability says at least two things: (1) the measures of stability and instability implicit in the simplified experience measure tap persistent or structural aspects of the U.S. labor market and (2) what changes the most over time is the marginal distribution of labor-force positions and possibly the association between positions and labor-market experience types.

Stability/Instability by Sex, Ethnicity, Age, Occupation, and Industry

Macro-level analyses of labor markets use a combination of demographic and structural factors in addition to the highly aggregated descriptions presented throughout this section. Preliminary findings based on this approach, however, indicate that our framework is helpful in elaborating the labor-force dynamics that differentiate workers in different groups or in different occupational or economic sectors. This can be seen clearly by considering simple marginal distributions of the labor-market-experience variable (trichotomized form) within demographic, ethnic, occupation, and industry groups. The percentages in the three main experience types in the 1981 recession year appear in Table 5.9.

Dramatic differentials are observed across race and ethnicity groups, and these are different by sex. Importantly, Hispanics are not located between nonblacks and blacks. For example, Hispanic men have a lower proportion in the stable inactive category (15.8%) than any other sex and race/ethnicity group,

Table 5.9
**Distribution of Prior Experience by Demographic Groups,
Occupations, and Industries (March 1982)**

Group	Stable Inactive (%)	Unstable Active (%)	Stable Full-Time (%)	Percentage in Category[a] (%)
Total population	30.6	31.5	37.8	100.0
Sex Group:				
Males	19.7	29.4	50.9	47.6
Females	40.6	33.4	26.0	52.4
Race/ethnic:[b]				
Nonblack	30.1	31.3	38.6	85.3
Males	18.9	28.8	52.3	[47.9]
Females	40.4	33.6	26.0	[52.1]
Black	34.6	32.8	32.6	10.7
Males	27.7	32.5	39.8	[44.9]
Females	40.2	33.1	26.8	[55.1]
Hispanic	32.2	32.2	35.6	3.9
Males	15.8	34.0	50.2	[48.3]
Females	47.5	30.6	21.9	[51.7]
Age				
16–19	34.3	60.0	5.7	9.2
20–34	15.0	40.3	44.7	34.6
35–49	16.4	27.6	56.0	22.3
50–64	33.3	21.8	44.9	19.2
65+	83.5	11.6	4.9	14.6
Occupation:[c]				
Professional	.0	31.9	68.1	10.6
Managers	.0	19.8	80.2	7.1
Sales	.0	49.8	50.2	4.4
Clerical	.0	46.4	53.6	12.5
Craft	.0	35.1	64.9	8.3
Operatives	.0	47.5	52.5	7.1
Transportation	.0	42.1	57.9	2.3
Nonfarm labor	.0	63.1	36.9	3.5
Housework	.0	86.6	13.4	.8
Other service	.0	64.8	35.2	9.3
Farmer/manager	.0	23.0	77.0	.8
Farm Labor	.0	70.3	29.7	1.0
No work experience	94.8	5.2	.0	31.8
Residual	100.0	.0	.0	.5
Industry:[d]				
Agriculture	.0	51.5	48.5	2.4
Mining	.0	37.1	62.9	.7
Construction	.0	54.1	46.0	4.3
Durable goods	.0	30.9	69.1	8.3
Nondurable goods	.0	36.8	63.2	5.7

Table 5.9. Continued.

Group	Stable Inactive (%)	Unstable Active (%)	Stable Full-Time (%)	Percentage in Category[a] (%)
Transportation/utility	.0	28.8	71.2	4.1
Wholesale	.0	29.4	70.6	2.8
Retail	.0	61.1	38.9	11.8
Finance	.0	31.5	68.5	3.9
Business repair	.0	47.1	52.9	3.0
Private household	.0	86.1	13.9	1.0
Personal	.0	58.6	41.4	1.8
Entertainment	.0	71.1	28.9	.9
Professional and service	.0	45.0	55.0	13.9
Public administration	.0	25.2	74.8	3.4
No work experience	94.8	5.2	.0	31.8
Residual	100.0	.0	.0	.5

[a]Values in brackets are percentages of the given race/ethnic group.
[b]Race/ethnic categories are nonblack nonhispanics, black nonhispanics, and Hispanics.
[c]1970 Census definitions of major occupational groups.
[d]1970 Census definition of major industry groups.

but their proportion unstable active is the highest. As might be expected, nonblack males are the least marginal group (stable active = 52.3%). Information in either experience measure might be used to elaborate race/ethnicity differentials in earnings, occupation, or other attainment measures.

Age variability can be characterized as follows. Inactivity is the highest at the lower and upper ages and the lowest in the prime labor-force ages, 20–49 (less than 20%). The proportion stable full-time active is the highest (56%) in the middle age category, 35–49, and is much lower at both the lower and the higher ages. And the unstable active category diminishes steadily across age groups, with the lower figures for the older age group denoting the combined influences of withdrawal from the labor force and stability in employment.

Occupation differentials reduce to a comparison of unstable active and stable full-time categories. This is because those who are inactive do not have occupations to report. The rank order of occupations in terms of the proportion stable active is as one would expect. Managers, farmers/managers, and professionals top the list, while nonfarm labor, farm labor, and housework are at the bottom. The relatively large magnitudes observed for the unstable active proportions are surprising. When we disregard the three most stable occupational groups, these proportions range from about one-third to over two-thirds. This is another indication of the pervasiveness of unstable or marginal labor-market experiences, and it should lead us to question the validity of assuming that incumbents of such occupations can all be assigned the same occupational status. Given the differences reported here, it is difficult to see how an explanation of occupational attainment could be justified without taking account of the labor-force dynamism involved.

Industry differentials reported in Table 5.9 are also dramatic. Stability is the highest in finance, public administration, durable goods, wholesale, and nondurable goods, which usually represent core-type industries. Instability is the highest in industries that represent the periphery sector.

SUMMARY: STRUCTURAL PERSISTENCE VERSUS CHANGE IN THE NEW MATRIX

We can summarize our conclusions about the relationship between prior work experience and current labor force position fairly easily;

1. *The distribution of experience types is reasonably stable over time for the total population and within major demographic groups.* However, there is a trend toward higher prevalence of the stable inactive category for males and a trend toward lower prevalence of this category for females.

2. *The trichotomous version of experience types is suitable for some inferences but not for others.* We can infer the magnitude of the association between experience types and labor-force positions from the trichotomous version but not the score parameters required for serious analysis of the latent quantitative variables associated with each facet of the contingency table.

3. *The marginal distribution of the trichotomous version of current labor force experience is essentially constant over time and hence "structural."* We need only disaggregate the middle category (unstable active) to examine the way that labor-market experiences change during periods of recession or growth.

4. *The association between experience types and labor-force outcomes is very strong and requires at least a "two-dimensional" representation.* Correlations are on the order of 0.8 or 0.9 on the first dimension and about 0.5 on the second dimension.

5. *There is a remarkable degree of temporal constancy in the association between experience types and labor-force outcomes,* so much so that this association may also be viewed as a structural or persistence feature of the labor-market matching process.

6. *Estimated scores for both experience types and position types seem to capture the major axes of row–column interaction.* The distinctive forms of association between particular experience types, or types of labor-market behavior, and particular labor-force positions, or types of un-

deremployment, can be retrieved from these score parameters. Importantly, these scores are based on a model recognizing as many categories of both variables as it is possible to measure given data limitations and the impossibility of measuring extremely rare types reliably. In addition, the score parameters are based on the total population, not just a segment of it.

SUMMARY AND DISCUSSION

We believe this framework provides a means to study labor-force dynamics that is automatically tied to most existing labor-force concepts, certainly to the vast majority of those concepts that have become the mainstays of official labor-force statistics. Reckoning with labor-force dynamics is important for labor-market research despite its relative neglect in contemporary sociology. We need to do much more to bring the forces of recession and growth, which are typically regarded in terms of labor-force concepts, into the sociology of labor markets. And we must recognize both the heterogeneity in labor-force behavior that actually exists and the dynamism in this behavior. When this is done, the instability of employment, worker marginality of various kinds, and transcience rather than persistence clearly emerge as central characteristics of the contemporary U.S. labor market.

We believe that this framework will also be useful as a means to incorporate labor-force behavior into other research traditions. A labor-force focus like the present one can inform studies of the underclass (see Lichter, 1988), research on ethnic differentials in economic attainment (Tienda, 1989), or research on poverty. There is certainly room for including better measures of stability, economic activity, and underemployment in all of these areas. Simple measures of current labor-force participation or unemployment are misleading because they mask the complexity of labor force activity involved and because they imply a temporal persistence that is inconsistent with the evidence. How can information from our framework be used in other settings where labor-force behavior is regarded as an important predictor or covariate? There are several possibilities that will be illustrated in the context of a conventional attainment model.

The prior experience by current position contingency table can be regarded as a block of covariates in most earnings or occupational status models. That is, labor-force behavior over the previous year as well as current labor-force position could be regarded as predictors or controls in those models. A simple approach for doing this is to include dummy variables for both key variables, or groupings of them. However, this would involve many terms and is probably unrealistic. The trichotomous version of current experience giving types that were called stable inactive, unstable active, and stable active would simplify analyses greatly. Other possible groupings of the categories have also been suggested above. The modeling strategies used here can also be exploited to considerable advantage. The score parameters from the RC(2) model, for example, can be used to capture most of the "variability" in this matrix. Using these score parameters in rather standard occupational attainment models is the topic of the next section.

One of the main advantages of our framework is that a much more inclusive sample (or universe) is used. Here, all civilian persons over age 16 were included. In contrast, many earnings-attainment models select only currently full-time employed workers with continuous work experience. The heterogeneity in labor-force behavior that we have taken such pains to measure would normally be dealt with indirectly in such models, either by including "corrections" for unmeasured heterogeneity or by using model-based adjustments for sample selection. An adjustment for sample selection bias (Heckman, 1979; Stolzenberg & Relles, 1990) is automatically built into a model that includes information from our prior experience × current position typology. We are optimistic about the potential of this framework for studying labor-force dynamics. But we expect benefits to be obtained bv using this approach to incorporate labor-force dynamics in more general settings where occupational status, earnings, sector placement, and the other key stratification variables are considered.

NOTES

1. For example, Kalleberg and Sørenson (1979, p. 351) define the labor market as an institutionalized mechanism by which "workers exchange their labor power in return for wages, status, and other job rewards." A more recent definition by Kalleberg and Berg (1987, p.48) is that labor markets "are arenas in which workers exchange their labor power, creative capacities, and even their loyalties with employers in return for wages, status, and other job rewards." The lack of attention paid to basic access issues is all the more salient in an economy where full-time/year-round work cannot be taken for granted (see Rifkin, 1995; Moore, 1996; *New York Times*, 1996; Dunkerley, 1996; reviewed in Leicht, 1998).

2. This exploratory analysis used information in some of the multiway contingency tables that can be formed from the items in the Appendix 5.1. In addition, the types we use were chosen so that reasonable marginal distributions were obtained and so that standard measures of past year's labor force behavior could be derived from the categories used. The definitions given here are consistent with all CPS files after 1976.

3. Tienda (1989) shows that the distinction between our stable inactive type and the more customary not-in-the-labor-force (NILF) category, a measure of current behavior, makes a big difference when one is analyzing difference between Puerto Ricans and other Hispanic groups. Results below highlight the differences between the relative persistence of the former and the relative absence of the latter.

4. The low-income category is based on work-related earnings over the previous year rather

than current income or income in the first two months of the current year. There is therefore some slippage in the "currency" of this measure. Unfortunately, this is the only kind of earnings measure available with CPS data. The income category will nevertheless be regarded as a measure of current (low) income status.

5. Using the imagery of status-attainment models, this contingency table intervenes between background factors and schooling on the one hand and occupational and earnings attainments on the other.

6. The cell frequencies are weighted with case-specific sampling weights that adjust for stratification and other complications. Case weights were adjusted so that the overall sample size ($N = 123,463$) is the same as the total of the unweighted frequencies, apart from rounding error. Weighted frequencies treated in the usual way provide consistent estimates of the parameters in log-linear and related models, but chi-square statistics measuring goodness of fit may be invalid (Clogg & Eliason, 1987). Fit statistics used below are descriptive indices of model-data agreement only. However, relative comparisons of them appear to be satisfactory for the types of inferences we are making here.

7. This fact emerges from inspection of marginal distributions that may be obscured by the above focus on transition or recruitment rates. The distribution among categories of the current position variable divides into about one-third (36.5%) not in the labor force, about one-third (32.2%) in adequate full-time employment, and about one-third (31.3%) in the current positions that represent current underemployment or marginal-worker status. The categories of prior experience can likewise be trichotomized into a nearly uniform distribution: stable inactive = 30.6%; categories 2–15, unstable active = 31.5%; and stable full-time active = 37.8%.

8. Although the prior experience by current position cross-classification resembles a mobility or turnover table, conventional models for mobility tables are not strictly appropriate for describing its properties. This is because the categories of prior experience are not entirely commensurate with the categories of current position. No simple regrouping of categories for either variable can enforce the one-to-one correspondence between row and column categories that formally characterizes most current models for occupational mobility tables. Both rows and columns represent multiple typologies derived from the simultaneous consideration of several different variables, implying that each variable might actually have more than one central dimension. Association models developed by Goodman (1984, 1986) and others (Clogg, 1982; Gilula & Haberman, 1988) for the more general context will be used here.

9. The full-year/full-time group corresponds to prior experience category 16 (score = – 0.48). The part-year/full-time group corresponds to prior experience categories 9–13 (scores = – 0.33, 0.67, 1.00, 0.40, 0.61, with mean = 0.47). The part-year/part-time group corresponds to prior experience categories 4-8 (scores = 1.81, 1.78, 1.82, 1.88, and 1.75, with mean = 1.75). And the full-year/part-time group corresponds to prior experience categories 14 and 15 (scores = 2.67 and 3.54, with mean = 3.10).

10. The ordering of current position categories from low to high is as follows: 1 (not in the labor force), 2 (unemployed), 7 (HV), 3 (UQ), 5 (HI), 4 (UL), 8 (I), 6 (HE), 10 (ADFT), 9 (M)

11. This graphic display is similar to the geometric representations used in correspondence analysis. As in correspondence analysis, we define rescaled scores as $\mu^*_{im} = (\phi_m)^{1/2}\mu_{im}$ and $v^*_{jm} = (\phi_m)^{1/2}v_{jm}$

12. This outlier is due almost entirely to the inclusion of men and women in the analysis. When women are removed from the sample, this point moves toward the center of the figure. E15 does not concern us because it refers to less than 1% of the sample.

13. Although not as extreme a case as E15, E9 (full time, part year, not looking) is a singleton. By inspecting the specific definitions of both prior experience categories (see Table 5.1 and the Appendix 5.1) and current categories, one will generally find that the matching of prior experience types with specific labor-force outcomes (current position categories) is very much

in line with earlier expectations. For example, group 1, consisting of E3 (nonworker, looked for 1–14 weeks), E2 (nonworker, looked for 15+ weeks), and P2 (unemployed, new entrants and reentrants), clearly constitutes a structurally similar set of categories with respect to both labor-market and labor-force behavior.

14. The rates for the unstable active category are weighted averages of the rates for the middle 14 categories of prior experience. They provide an aggregate summary that is difficult to predict from the inflow or outflow rates based on prior experience. The unstable active category is associated with a substantial rate of nonparticipation in March 1982 (23.6%).

15. The marginal distribution of our simplified prior experience measure is radically different between the sexes: 20, 29, and 51% for males, compared with 41, 33, and 26% for females. While the relative magnitude of the unstable active category is practically the same for males and females, the other two categories have quite different relative frequencies, with the stable inactive category about twice as prevalent for females as for males. Note that this category is a much more restrictive definition of inactivity than the usual measure of labor-force participation. If we restrict attention to just those who are "labor-force eligible" (experience categories = 2 or 3), the unstable category is even more salient for females: 56% versus 37% for males.

16. The full information for creating the prior experience variable is available only for CPS files since 1977, so we restrict attention to years 1977–85, a 9-year series that exhibits considerable labor-market and labor-force variability. The 1982 data used for all analyses above provide a benchmark for assessing variability because the 1981–82 period was the worst recession since the Great Depression

17. We can also show that the association between experience types, expressed in the simple trichotomy, and labor-force outcomes is remarkably stable over time. We looked at correlations between experience and current position scores based on the RC(2) model applied to the nine CPS years 1977–85, for each of five different demographic groups: (1) the total population, (2) males, (3) females, (4) ages 16–19, and (5) ages 20–34. (Results for other age groups are similar and so will not be reported.). For the total population, the first-dimension correlations range from about 0.77 to about 0.79, and the second-dimension correlations range from about 0.39 to about 0.44. Given the remarkable degree of temporal constancy, the association can be understood as a structural feature of the relationship between experience types and labor-force outcomes. This temporal constancy does not change with the demographic group considered (see Clogg, Eliason, & Wahl, 1990, Table 12).

APPENDIX 5.1. VARIABLES IN THE CURRENT POPULATION SURVEY USED TO MEASURE CATEGORIES OF LABOR-MARKET EXPERIENCES

The March CPS contains a rather detailed list of items that can be used to classify workers and nonworkers according to job-search behavior or empirical consequences of that behavior. The following information is available for CPS files since 1977. Earlier files contain only some of the relevant items and so could not be used to obtain the full typology used in this paper.

Item 1: Weeks looking all in one stretch (STRETCH)
>   0. Not in universe (NIU)
>   1. 1 week
>   2. 2 weeks
>   3. 3 or more weeks

Item 2: Reason worked part time (RSPART)
>   0. NIU
>   1. Could find only part-time work
>   2. Wanted or could only work part time
>   3. Slack work or material shortage
>   4. Other

Item 3: Part–full-time recode (PTFTREC)
>   0. NIU
>   1. Full time (full-year worker)
>   2. Part time (full-year worker)
>   3. Full time (part-year worker)
>   4. Part time (part-year worker)
>   5. Nonworker

Item 4: Work recode 1; work experience, weeks looking–nonworkers (WRKREC 1)
>   0. NIU
>   1. None (not looking for work)
>   2. 1–4 weeks looking
>   3. 5–14 weeks looking
>   4. 15–24 weeks looking
>   5. 25–39 weeks looking
>   6. 40 or more weeks looking

Item 5: Work recode 2; part-year worker, weeks looking (WRKREC2)
>   0. NIU
>   1. None
>   2. 1–4 weeks
>   3. 5–10 weeks
>   4. 11–14 weeks
>   5. 15–26 weeks
>   6. 27–39 weeks
>   7. 40 or more weeks
>   8. Full-year worker
>   9. Nonworker

Specific categories obtained from these five items define the 16 types in the EXPER variable in Table 5.1.

APPENDIX 5.2. ADDITIONAL NOTES ON MODEL ESTIMATION

The model used here assumes that row categories mark relative positions on a quantitative but latent "experience" variable. And it assumes that column categories mark relative positions on a quantitative but latent "labor-force position" variable. The cross-classification of EXPER and POSITION will be used to infer metric properties of both variables. Relationships between the two latent variables, summarized in the score parameters for the model, will be taken as a representation of the labor-market matching process.[1]

Let F_{ij} denote the expected frequency in cell (i, j) of the two-way table, and let $P_i = F_{i+}/N$ and $P_j = F_{+j}/N$ denote the row and column marginal distributions, respectively. The RC(M) association model in Goodman (1986) or Becker and Clogg (1989) can be written as follows:

$$F_{ij} = \eta \alpha_i \beta_j \exp\left(\sum_{m=1}^{M} \phi_m \mu_{mi} \nu_{mj}\right) \tag{5.1}$$

The μ and ν parameters represent scores for row and column categories, respectively. We say that μ_{mi} is the score, in the m-th dimension, for row category i and that ν_{mj} is the score, in the m-th dimension, for column category j. The intrinsic association in the m-th dimension is ϕ_m for $m = 1, \ldots, M$. The score parameters are standardized as follows:

$$\sum_{i=1}^{I} P_{i.} \mu_{mi} = \sum_{j=1}^{J} P_{.j} \nu_{mj} = 0 \qquad \text{(scores have mean zero)} \tag{5.2a}$$

$$\sum_{i=1}^{I} P_{i.} \left(\mu_{mi}\right)^2 = \sum_{j=1}^{J} P_{.j} \left(\nu_{mj}\right)^2 = 1 \qquad \text{(scores have variance one)} \tag{5.2b}$$

For purposes of identifying scores, the cross-dimension correlation of m_m and $m_{m'}$ and the cross-dimension correlation of ν_m and $\nu_{m'}$, for $m \neq m'$, are set at zero (orthogonal scores). The meaning of the parameters is further clarified by considering how the model decomposes local odds ratios.

---

1. If panel data were available, $EXPER_t$ and $EXPER_{t-1}$ might be cross-classified to scale levels of the EXPER variable. But then the parameters would not tell us how current POSITION levels are related to EXPER levels. Similarly, we might use $POSITION_t$ and $POSITION_{t-1}$ to scale levels of the POSITION variable, but this would not include any information on a labor-force behavior throughout the year. The cross-classification of EXPER and POSITION provides a direct linkage between a previous year's labor-force behavior and the current labor-force position. The matching process implicit in this table is exactly what we wish to summarize through the use of these special models.

Let $\Phi_{ij(i'j')}$ denote the logarithm of the odds ratio obtained from the 2×2 table obtained by taking rows i and i' and columns j and j':

$$\Phi_{ij(i'j')} = \log\left[\left(F_{ij}F_{i'j'}\right)/\left(F_{i'j}F_{ij'}\right)\right] \tag{5.3}$$

With the model in Eq. (5.1), these quantities can be written as

$$\Phi_{ij(i'j')} = \sum_{m=1}^{M} \phi_m \left(\mu_{mi} - \mu_{mi'}\right)\left(v_{mj} - v_{mj'}\right) \tag{5.4}$$

Equation (5.4) says that the association in the particular 2×2 table, as measured by the log-odds ratio, can be decomposed into M components or "dimensions" each corresponding to a set of row and column scores. In the first dimension, the intrinsic asssociation is ϕ_1, which can now be interpreted as the log-odds ratio when both the row scores and the column scores are one unit apart. Given the restrictions on the scores in Eqs. (5.2a) and (5.2b), ϕ_1 is thus the value of the log-odds ratio for row contrasts and column contrasts that are 1 SD apart on the metrics implicit for both sets of scores. This intrinsic association is multiplied by the distance between rows i and i' (i.e., μ_{1i} – $\mu_{1i'}$) and the distance between columns j and j' (i.e., v_{1j} – $v_{1j'}$) to obtain the contribution to local association from the first dimension. Similar comments apply to the scores and intrinsic association parameters pertaining to the other components or dimensions. As in Goodman (1986), this model can be used to summarize the dependence of columns on rows, with the row classification as a predictor of columns. This model may be viewed as a description of outflow rates, inflow rates, or the joint distribution.

APPENDIX 5.3. DISAGGREGATION OF THE 1981–82 ASSOCIATION BY SEX AND AGE

It is natural to consider both sex and age differentials in the EXPER × POSITION table, or in model parameters for this table, further to refine and test interpretations offered above. Here we consider issues of relative fit of the model and measures of intrinsic association. Score parameters differ, sometimes substantially, by sex or by age, and these are not examined closely here. However, the score parameters as metrics for the entire population should be based on the table containing all ages and both sexes, so differentials in score parameters by sex or age should not be viewed as a threat to the validity of the framework.

We examined chi-squared values for the RC(0), RC(1), and RC(2) model;

applied separately to the sex groups and separately to four age groups. In each case, the RC(2) model accounts for at least 90% of the baseline value; this model appears to be satisfactory for all groups. The intrinsic association between EXPER and POSITION is slightly larger for females (ϕ = 1.50) than for males (ϕ = 1.32). This appears to be due to a greater "persistence" in economic inactivity for females than for males (i.e., a high rate of flow from NW-NL on EXPER to NILF category on POSITION), which is a special kind of row–column interaction (matching).

The pattern of intrinsic association by age group reveals essential similarity between age groups 20–34 and 35–49 but much less association (i.e., movement that is closer to random sorting) for ages 16–19. There is a monotonic increase in the magnitude of the intrinsic association with age, with first-dimension correlations ranging from 0.52 for the teenage group to 0.88 for the 50–64 group. In other words, there is more "structure," or greater association, in the relationship between labor-force behavior and labor-force position in the older age groups, which is the pattern expected.

APPENDIX 5.4. LOSS OF STRUCTURAL INFORMATION BY CONDENSING

To what extent can collapsing categories of EXPER be defended on statistical grounds? To answer this question, we will compare some inferences based on EXPER3 with the corresponding inferences based on EXPER. We examined the chi-square values and some other quantities for models applied to the 3 × 10 cross-classification of EXPER3 with POSITION. Because the RC(0) model is the model of row–column independence, the L^2 values for the 16 × 10 and the 3 × 10 tables can be directly compared to examine the contribution due to nonindependence in the middle 14 categories (Goodman, 1981b). From Table 5.5, the L^2 value for the RC(0) model is 146,522; from Table 11, the L^2 value for the same model with collapsed categories is 123,674. The difference between these two values is 22,848, the chi-square component for nonindependence in the middle 14 categories of EXPER. This is a 15.6% reduction in the fit statistic, a moderate loss of information when the 16 × 10 table is replaced by the 3 × 10 table. On the other hand, there are indications that the structure of the association is duplicated to a remarkable extent after condensing.

To examine the so-called structural criteria for collapsing, consider first the intrinsic association between rows and columns as measured by either the f parameters or the correlations between scores. For the total population, f_1 and f_2 have estimated values of 1.585 and 0.446 in the 3 × 10 table compared with the corresponding values of 1.583 and 0.404 in the 16 × 10 table (see

Table 5.6). In other words, the intrinsic association is about the same in both the uncollapsed and the collapsed versions of the data. The only difference is a slight increase in the association in the second dimension. For the total population in 1981–82, the extreme condensation of categories used to produce the EXPER3 variable seems reasonably valid, especially given the benefits of using the simpler variable in the types of analyses given above. For males the reduction in intrinsic association is also modest, ϕ_1 = 1.261 in the condensed table versus 1.318 in the full table. For females, the intrinsic association is slightly greater in the collapsed table, ϕ_1 1.627 versus 1.503. Note that L^2 values for the RC(1) and RC(2) models are not directly comparable across tables; for example, the RC(2) model is saturated for the 3 x 10 table, implying that $L^2[RC(2)] = 0$.

An alternative way to examine the loss of information resulting from condensing categories is to compare estimates of category scores from the RC(2) model for the two kinds of tables. The scores for the RC(2) model using the EXPER × POSITION table were presented in Table 5.6. EXPER3 scores on the first dimension [estimated from the RC(2) model applied to the 3 × 10 table] were $\mu_{1,1}$ = –1.44 (compare with –1.28), $\mu_{1,2}$ = 0.26 (compare with middle μ_{1i} values in Table 5.6), and $\mu_{1,3}$ = 0.95 (compare with 1.07). The POSITION scores on the first dimension change rather substantially when the 3 × 10 table is used in place of the 16 × 10 table. The first-dimension scores for the 10 categories of POSITION are –1.17, –0.41, 0.41, 0.83, 0.31, 0.74, 0.33, 2.44, 0.10, 0.74 (v_{1j}, j = 1, . . . , 10). These should be compared with the first-dimension scores for POSITION (v_{1j}, values) in Table 5.6. The switching of order among the last several of these scores is perplexing. Although we can make inferences about the strength of association fairly well with the condensed table, it is clear that the 16-fold classification of labor-market experience types cannot be replaced by the cruder classification if the goal is to scale POSITION categories. In sum, using EXPER3 leads to modest loss of information when a chi-square criterion is used (about 15%), but some information is maintained when structural criteria (model parameters) are used. That is, inference about the intrinsic association is essentially unchanged, the score parameters for EXPER3 are still reasonable, but the score parameters for levels of POSITION change substantially. Inferences involving score parameters should be based on the model for the uncollaped version of the table, but inferences about the intrinsic association can be based on the collapsed version of the table without serious risk of bias.

6

Labor-Force Behavior
and Its Influence on Status
and Wage Attainments

INTRODUCTION

 By now, the case for taking into account labor force activity when analyzing economic opportunity should be clearly established. To review our findings so far:

1. At any given period since 1970, 15–25% of the labor force is marginally employed (or underemployed), using relatively conservative definitions.
2. Between 30 and 40% of the labor force at any time has unstable or marginal labor market experiences. These experiences can be observed by looking at labor force activity over time periods as short as 1 year.
3. The restructuring of jobs, occupations, and industries, along with effects of economic recession and growth, have produced marked fluctuation in labor force categories that ought to be reconciled with essentially static views of a system of occupational positions.

   In most prior status attainment research, marginal labor force classifications, unstable labor market experiences or behavior, and economic fluctuation in the labor force have been completely unexamined.[1] In this section we provide a way to measure complex patterns of labor force behavior that can be applied using many existing data sets. Our goal here is to summarize labor force activity with a few quantitative measures that sufficiently reflect the large numbers of qualitative categories such as unemployed, underemployed, adequately employed, working poverty, and so on. We show how these new variables work in the context of a simple model of occupational attainment applied to 1982 CPS data that describe labor force changes during the worst recession since the 1930s.

In the next section, we define measures of *labor-force behavior* (defined in terms of labor market experiences) and *labor force positions* that can be replicated or closely approximated with most large databases used in stratification research. A cross-classification of prior behavior by current position is then analyzed using an association model that scales both factors. The model applies to the civilian population over age 16.[2] Score parameters derived from this model can be viewed as intervening factors in conventional status attainment models or their modern relatives. Finally, the score parameters that capture the main features of the behavior-by-position contingency are used as covariates in respecified attainment models and are shown to be substantively and statistically important. We find that these new covariates have very large effects on occupational attainment, which argues for a serious consideration of our framework.

LABOR FORCE MEASURES ADDED TO OCCUPATIONAL ATTAINMENT MODELS

The 1982 CPS data were used to estimate a single-equation model of occupational attainment. We use this single-equation model as a test case. We are interested primarily in the effects of these new variables rather than the interpretation of effects of other parts of our simplified analysis. The Duncan SEI is used as the measure of occupational status as in most previous studies. CPS data do not provide information about background characteristics such as father's occupation, but because most of the effect of background disappears once schooling is included, we do not view this omission as a serious deficiency. Direct information on firm characteristics or authority structures is also absent. These and other limitations restrict the degree to which the best current models of occupational attainment can be used. To highlight the utility of our modifications we consider three simple regression equations for samples of men and women.

Men

Table 6.1 gives results for men in the 1982 CPS. Model Equation (1) is intended to represent a highly restrictive sample selection criterion; it pertains to the currently full-time employed population ages 25–64, the type of

Table 6.1
Occupational Attainment (Duncan SEI Scores):
Three Models for Men

Predictor	Model I	Model II	Model III
Constant	−.040	−3.332	2.365
	(3.200)	(2.048)	(2.014)
COLOR[a]	−8.198	−7.569	−9.380
	(3.362)	(2.165)	(2.123)
AGE	.091	.165	.083
	(.040)	(.023)	(.023)
COLOR*AGE	.160	.146	.174
	(.042)	(.146)	(.024)
GRADE[B]	1.779	1.611	1.272
	(.211)	(.146)	(.144)
COLOR*GRADE	.900	.851	.878
	(.224)	(.156)	(.106)
COLLEGE[B]	6.514	6.648	6.415
	(.250)	(.204)	(.200)
COLOR*COLLEGE	−.329	−.224	−.185
	(.257)	(.210)	(.206)
MSA[c]	3.346	3.359	3.322
	(.241)	(.186)	(.182)
EXPER–1	—	—	2.782
			(.197)
EXPER–2	—	—	−.668
			(.125)
POSITION–1	—	—	−.071
			(.227)
POSITION–2	—	—	−2.312
			(.116)
R^2	.413	.421	.444
N	28,491	45,404	45,404
Root MSE	18.762	18.524	18.156

Note: Standard errors (assuming simple random sampling with case weights incorporated) in parentheses. EXPER–1, EXPER–2, POSITION–1, and POSITION–2 are estimated scores from RC(2) model from Table 5.6. Model I is based on currently full-time employed persons 25–64. Models II and III are based on all people with occupations.
[a]COLOR coded 1 for nonblacks, 0 for blacks.
[b]GRADE and COLLEGE are spline-coded variables (completed years of SCHOOL = GRADE + COLLEGE).
[c]MSA coded 1 if household is in an MSA, 0 otherwise.
Taken from Clifford C. Clogg & Scott R. Eliason, 1990, "The Relationship between Labor Force Behavior and Occupational Attainment." *Research in Social Stratification and Mobility*, Vol. 9, p. 173.

sample usually considered for studying earnings attainment. Note that this sample selects only 63% (28,491/45,404) of the population over age 16 with an occupation. Variables included in the regression are race (1 = nonblack, 0 = black), age, race*age, completed years of schooling coded as a spline function (grade and college), race*grade, race*college, and SMSA (1 = resides in SMSA, 0 otherwise).

Equation (2) includes the same predictors but uses the entire sample over age 16 with a reported occupation. Results from Models Eq. (1) and (2) are remarkably similar with the possible exception of the age coefficient [0.091 in Model Eq. (1), 0.165 in Model Eq. (2)]. Model Equation (2) will be used as a baseline for comparing a conventional status attainment model with a respecified model including measures that summarize labor force behavior.[3]

Model (3) adds to Model Eq. (2) the prior experience and current position scores we constructed earlier, which are metric variables summarizing the relative position of labor force members on the two latent dimensions of prior experience and current position. Note that this model, like the previous ones, selects only those with reported occupations, which in the CPS means that only those currently in the labor force were considered. On the other hand, the score parameters used here as metric covariates were estimated with the *entire civilian population*, so the score values included in this model adjust for sample selection in an explicit way because those without a current position in the labor force affect the position and experience scores of those who do.[4]

Adding the prior experience scores and the current position scores to the baseline status attainment model changes the inferences in a modest but important way. The R^2 increases by 2.3% (to 44.4%), and the standard error of the estimate decreases by over one-third of a unit (to 18.16). Scores on the first dimension of prior experience, which reflect relative stability in nonmarginal employment, have a dramatic effect on the SEI (coefficient = 2.78, t = 14.1). Stability in employment or labor force behavior is strongly and positively related to occupational status.[5]

Scores on the second dimension of prior experience, which essentially contrast full-year versus part-year work and full-time versus part-time work over the previous year, have a substantial negative effect on the SEI (-0.67, t = -5.3). Note that these scores are negative for full-time or full-year work and positive for part-time or part-year work, so the negative sign of the relationship is anticipated. In others words, full-year and/or full-time employment histories are positively associated with occupational status.

To interpret the magnitude of these effects, it is important to recognize that the scores were identified by setting the standard deviation equal to unity. Looking at the other estimated values and taking account of the metrics involved, the coefficient of the prior experience-1 scores appears to be at least as important as the age effect, for example, although it is certainly not as important as the effect of schooling. A standard deviation increase in prior experience-1 (EXPER-1) scores produces an increase of 2.8 Duncan SEI units, which is about the same change expected from a $2.782/0.083 = 33.5$-year increase in age, or a $2.782/1.272 = 2.2$-year increase in years of graded schooling, or a 3-month increase in college education. Viewed in this light, the effects of the prior experience scores are fairly dramatic.

Current labor force position also affects (or is associated with) occupational status, although only the second-dimension scores are important in statistical terms. The second-dimension scores appear to be quite important. These scores essentially define a contrast between economic underemployment (including unemployment and part-time employment) and full-time employment with large positive values assigned to the former group. (Note that current position category 1, representing the not-in-labor-force status, is excluded because the sample includes only those currently in the labor force.) A negative effect would be predicted, and this effect is large (-2.3) and very significant ($t = -19.9$). A full-time job is associated with a substantially higher SEI value, net of other effects in the model. Although the sign of the relationship is as expected we are surprised that the effect is so strong. In contrast, the vertical dimension of current labor force position (differentiating those currently in and out of the labor force) has little effect when the other behavior or position measures are included. In summary, including both sets of scores for labor market experiences and the second-dimension scores for current labor force position adds important effects to the model of attainment.

Prior labor-force experience and current position clearly affect socioeconomic status in our analysis. But does controlling for labor force status alter the effects of the other characteristics? The simple answer is that adjusting for prior experience and current position makes the most difference for blacks, the group most marginalized in the labor market. Specifically,

1. *The age effect for blacks is one-half as large in the analysis adjusting for labor force experience and position*, and the age effect for others is 17% less. This reflects the fact that labor-force experience and current position vary by age (see Chapter 3) and there are substantial racial differences in labor force participation across age groups.

2. *The education effect for blacks is lower when labor force experience and current position are taken into account* (by approximately 20%). We do not find this effect for other racial groups. Hence, much of the effects of education for blacks occurs because education moves blacks across experience and position categories, rather than improving their relative standing *within* labor force categories. For example, educational achievement may move blacks from unemployment or part-time employment to full-time/full-year employment. For other racial groups, changes occur *within* the full-time/full-year group as one moves from a less desirable to more desirable full-time position.

3. *Education also increases educational mismatch for blacks*, and controlling for mismatch would lessen the education effect on eventual socioeconomic attainment.

Women

Next, the results for women are presented in Table 6.2. Models (1) and (2) are virtually identical in terms of R^2 (27%). But note that the sample selection criteria for Model Eq. (1) are dramatic: only 53% (19,148/36,023) of all women with occupations are included when we select women aged 25–64 who are employed full time. The meaning of occupational status or occupational mobility for women as a whole is unclear (see Hout, 1988). About 20%

Table 6.2
Occupational Attainment (Duncan SEI Scores):
Three Models for Women

Predictor	Model I	Model II	Model III
Constant	–5.974	–.126	6.150
	(4.000)	(2.449)	(2.402)
COLOR[a]	–4.807	–12.276	–14.324
	(4.272)	(2.618)	(2.563)
AGE	–.128	–.098	–.143
	(.040)	(.024)	(.023)
COLOR*AGE	.136	.192	.208
	(.042)	(.025)	(.024)
GRADE[B]	3.541	2.581	2.172
	(.290)	(.182)	(.179)
COLOR*GRADE	.768	1.242	1.419
	(.313)	(.197)	(.193)
COLLEGE[B]	5.898	6.304	6.006
	(.238)	(.200)	(.196)
COLOR*COLLEGE	–2.150	–2.076	–1.987
	(.248)	(.209)	(.204)
MSA[c]	3.179	4.287	3.924
	(.282)	(.206)	(.202)
EXPER–1	—	—	1.622
			(.164)
EXPER–2	—	—	–.524
			(.099)
POSITION–1	—	—	.911
			(.227)
POSITION–2	—	—	–1.762
			(.098)
R^2	.270	.270	.301
N	19,148	36,023	36,023
Root MSE	17.865	18.298	17.903

Note: See Table 6.1 for description of quantities.
Taken from Clifford C. Clogg & Scott R. Eliason, 1990, "The Relationship between Labor Force Behavior and Occupational Attainment." *Research in Social Stratification and Mobility*, Vol. 9, p. 176.

of the female labor force is engaged in part-time employment, most of which is recorded as voluntary part-time work, and a substantial fraction of the remainder are underemployed. Our adjustment for labor force behavior works in the direction of making occupational status (controlled for current position scores) more comparable between men and women.[6]

The model with prior experience and current position changes the inferences one would make about positions in the stratification system. Worker stability is positively associated with occupational status and part-time status in prior and current positions reduces current occupational status substantially. By comparing men's and women's results, one would conclude that current position is much more important for women than men and that prior experience is more important for men than women.

The effects of the other measures change as well. Age effects are enhanced for women. The effects of schooling beyond high school is reduced for black women, but increases for other women and the college effect declines for black women as well. The major implications of these results are the same for women and men: Labor-force position and prior experience alter the effects of education and age for those traditionally marginalized in the labor market. Age and education help marginalized groups to move *between* labor force categories. Nonmarginalized groups (whites and men) gain in occupational status *within* labor force categories. The implications of this analysis would suggest that reducing the relative disadvantage of marginalized groups would begin by finding or creating stable, full-time/full-year jobs. Then policy analysts could focus their efforts on reducing inequality among those full-time/full-year positions.[7]

Summarizing Our Results So Far

Accounting for marginal and unstable labor force activity is a key component of understanding social stratification in the United States. Stratification researchers have rarely considered these issues as central. The analysis so far would suggest that the role of these labor force factors in shaping the life chances of individuals or groups should be considered central in our models and theories of stratification, and in subsequent policy recommendations. In this section we have tried to show that we can measure a complex set of characteristics with relatively simple quantitative scales that serve to reflect labor force factors that can easily be included in our models (see also Clogg, Eliason, & Wahl, 1990). Moreover, we provided evidence that these labor force factors have a strong relationship with occupational status and that they change inferences about effects of schooling and other factors in nontrivial ways, particularly for groups like women and blacks, who are marginalized and underemployed in the labor market.

This approach can be applied to more general specifications of status or wage attainment models in current use, including multiple-equation models with multiple indicators of key constructs. With longitudinal or panel data, for example, it would be straightforward to analyze a more complete recursive chain of influence, perhaps with past occupation viewed as a predictor of labor force behavior and/or underemployment. The strategy of combining several standard measures of labor force behavior into a set of a few quantitative covariates appears to be an attractive method for reducing the complexity of the problem. We believe that it would be profitable to use this approach to include labor force behavior in most current models of the attainment process, including models for earnings.

ADDING NEW SCALES TO PRIOR ANALYSES
OF LABOR MARKET STRATIFICATION

In this section we explore further the effects of labor force behavior on standard stratification outcomes—earnings and occupational status. As we discussed previously, nearly 40% of the labor force is underemployed during any year. Past research shows that employers take into account labor force careers in addition to human capital and ascribed characteristics (such as gender and race) in allocating rewards and that they evaluate recent labor-market experiences as they attempt to screen out workers with low productivity potential (see Althauser & Kalleberg, 1990; Wilson, 1996).

Adding the New Scales to Standard Attainment Models

Our point of reference is the basic attainment model reported in Table 5.6 of Featherman and Hauser (1978, p. 235). As in that source, our analysis uses the OCG-II data and all analyses are restricted to men aged 25–64 in the experienced civilian labor force, which implies that occupation has been coded for all cases included.[8] The outcome variables are occupational status, as measured by the Duncan Socio-Economic Index (SEI), and the natural logarithm of work-related earnings in the previous year. The standard predictors used to form baseline models are drawn from Featherman and Hauser (1978), which essentially replicates the work of Blau and Duncan (1967). These are *father's occupational status, father's level of schooling, number of siblings, farm origin, broken families, race, respondent's level of schooling,* a standard proxy for years of *labor-force experience* (Age - Years of Schooling - 6),[9] *labor force experience squared,* and the *occupational status* of the respondent's first job.

Occupational Status

Table 6.3 gives results from five models that predict the Duncan SEI. The same cases were used to estimate all models, so standard assumptions about missing value exclusions are implicit (see Little & Rubin, 1987). Because our measures of prior experience and current position are completely exhaustive, our labor force activity scales do not bias the results by deleting additional cases.

H_1 is the basic model including background factors but excluding schooling and first-job status, while H_5 is the basic model including schooling and first-job status in addition to background.[10] H_2 adds the four labor force scales derived from the analysis of prior experience and current labor force position, and described above. In this model, all four scales have large effects, but the

Table 6.3
Occupational Attainment (Duncan SEI Scores) Predicted from Standard Models and Models Containing Labor Force Scales (Men, Age 25-64)

Predictor		H_1 [a]	H_2	H_3	H_4	H_5
Background:	FOCC	.224	.234	.133	.065	.062
	FEDIC	.869	.852	_-.006_	_-.001_	_-.009_
	SIBLING	-.968	-.925	-.163	-.083	-.094
	FORIG	-3.935	-3.335	-.595	-.447	-.575
	BROKENF	-3.657	-3.355	-.595	-.447	-.575
	BLACK	-8.511	-7.802	-5.465	-5.024	-5.447
Schooling:	REDUC			4.016	2.342	2.416
First Job	DUNCF				.372	.374
			Labor force behavior scales			
	E1		4.855	4.385	3.894	
	E2		-4.268	-2.482	-2.551	
	P1		2.494	1.382	_.881_	
	P2		_.619_	_.223_	_-.066_	
Constant		33.214	25.341	-19.922	-7.627	-3.688
R^2		.197	.218	.411	.488	.479

Note: See text for a definition of variables. Insignificant coefficients (2-tailed test at .05 level) underlined. N of cases (unweighted number) is 19,012; there are no "missing values" in labor force scales. Standard (listwise) defaults on missing value exclusions were applied. The model predicting the Duncan SEI with the labor force behavior scales alone (results not shown) gave R^2 = .028 (P1 not significant). Note also that age is not included in the models (Featherman and Hauser considered separate analyses by age groups using these basic models).

[a]H1 is the baseline model in Table 5.6 in Featherman and Hauser (1978, p. 235). Our coefficients differ slightly from theirs (.247, .886, –1.182, –4.825, –2.470, –8.649, and [constant] 33.32, with R^2 = .206). We have been unable to account for the slight discrepancies, although differences in the use of sample weights (SPSS^x defaults used here) might explain them. Also, Featherman and Hauser used pairwise deletion of missing cases, not listwise deletion.

Taken from Clifford C. Clogg, Nimpha Ogena, & Hee-Choon Shin, 1991, "Labor Force Behavior in the Process of Socioeconomic Attainment: New Scales Added to Classical Models." *Social Science Research*, Vol. 20, p. 264.

proportion of variance explained increases only modestly (about 2%). H_3 adds to H_2 the years-of-schooling variable. Adding respondent's education to the model increases explained variance by 19%. H_4 adds status of first job; comparing H_4 and H_5 shows that the measures of current position do not have significant effects on current status once respondent's education is controlled. The inference is that measures of current labor force (or underemployment) status need not be considered once the other predictors are included.

On the other hand, the two experience measures continue to have strong effects in the most comprehensive models considered. These summary measures of prior labor force activity are strongly related to occupational status net of background, schooling level, and first job status, implying that recent labor market experiences exert independent effects on current occupational attainment. Whether the behaviors captured in the experience measures sort workers in a selective manner (reflecting a supply-side constraint) or whether employers allocate occupational rewards using worker characteristics captured in the experience measures as screening devices (reflecting a demand-side constraint) is something that cannot be answered with these data.

Comparisons with schooling effects are useful to acquire a sense of how important the effects of labor force behavior are. In the OCG-II sample, the standard deviation of Prior experience-1 is 0.59.[11] A change on the first dimension of prior experience of this amount is almost equivalent (0.59×3.894 = 2.30) to the effect of 1 year of schooling (2.34). The maximum contrast possible for this variable is about 2.75 (compare the score for level 3 of prior experience with the score for level 16), and such a contrast would lead to a "net" difference in occupational attainment of about 11 units on the Duncan SEI. This is about the same as a 4.5 year contrast in years of schooling. Although the increment to explanatory power (R^2) is modest, the effects of prior experience are substantial as judged by comparisons with schooling effects.[12]

Log-Earnings

We look at a more exhaustive list of models for log-earnings in Table 6.4. H_1 is the baseline model including only background factors and ascribed characteristics. Adding the new scales to this model gives H_2, and the consequence is dramatic: R^2 is increased from about 8% to over 26%. Models H_3–H_5 add respondent's education, the proxy for years of labor force experience, and its squared term in succession, giving increments to explanatory power comparable with those reported by Featherman and Hauser (1978). But note that *the coefficient values for the prior experience and current position measures are essentially unchanged with the inclusion of schooling and experience.* This adds credibility to our claim that the new scales add some new characteristic to the analysis, or that the influences are indeed independent or incremental effects.

Table 6.4

Log-Earnings Predicted from Standard Models and

Models Containing Labor Force Scales (Men, Age 25–64)

Predictor	$H_1{}^a$	H_2	H_3	H_4	H_5	H_6	H_7	H_8	H_9
				Background					
FOCC	.002	.003	.001	.001	.001	.001	.001	.001	.001
FEDUC	.012	.011	−.001	.002	.004	.004	.006	.006	.004
SIBLING	−.016	−.013	−.003	−.004	−.004	−.005	−.004	−.002	−.004
FORIG	−.126	−.129	−.084	−.103	−.091	−.082	.050	.065	−.090
BROKENF	−.085	−.068	−.030	−.027	−.032	−.043	−.053	−.043	−.033
BLACK	−.318	−.269	−.241	−.222	−.220	−.255	−.354	−.337	−.221
			Schooling and experience						
REDUC			.055	.064	.061	.090	.070	.061	.061
X				.006	.033	.044	.050	.041	.034
X^2					−.001	−.001	−.001	−.001	−.001
			Labor force behavior scales						
E1		.390	.398	.382	.368			.369	.350
E2		−.343	−.314	−.315	−.308			−.308	−.331
P1		.191	.201	.202	.182			.180	
P2		−.044	−.067	−.073	−.058			−.057	
λ							−6.951	−8.192	
Constant	2.198	1.690	1.074	.819	.598	.826	.848	.623	.591
R^2	.079	.261	.308	.318	.332	.183	.183	.333	.331

Note: Notes for Table 6.2 and Table 6.3 apply here as well. Cases with negative or zero earnings are treated as missing. N of (unweighted) cases is 20,741.

[a] H_1 is the baseline from Featherman and Hauser (1978, pg. 235). They report an R^2 of .027 (we think .072 should have been reported), and our results differ somewhat from those they reported for this model. The model predicting log-income from the labor force behavior scales alone (results not shown) gave $R^2 = .243$ (all coefficients significant).

Taken from Clifford C. Clogg, Nimpha Ogena, & Hee-Choon Shin, 1991, "Labor Force Behavior in the Process of Socioeconomic Attainment: New Scales Added to Classical Models." *Social Science Research,* Vol. 20, p. 266.

H_6 is the standard Featherman–Hauser model including background, schooling, and experience, with an R^2 value of 18%. Note that the model with the labor force scales added to this set (model H_5) leads to a substantial increment to explanatory power, with the majority of the effect obviously attributable to the measures of past year's labor force behavior. Note that with this model the signs are all in the expected direction: positive for vertical dimension of past behavior, negative for the special contrast represented by the second dimension of past behavior, positive for the vertical dimension of current position, and negative for the special contrast represented by the second dimension of current position (high positive values for underemployed workers, negative values for full-time employed persons, which should yield a negative relationship). The increment to explanatory power is impressive. The new scales almost double the R^2 value (increase from 18 to over 33%). We now consider two other specifications so as to examine some objections that might be made about the models presented thus far.

The possibility of sample selection bias is one obvious factor that ought to be examined. Heckman's (1979) method was used to check for this. The market experience measures were designed, to some extent at least, to characterize types of workers that might be treated as missing in conventional earnings attainment models. As such, including the new scales for these types can be viewed as a partial adjustment for selection effects. We now estimate models that attempt to control for selection bias without using the information in our new scales. A probit selection equation was estimated from the existing set of independent variables except the labor force behavior scales, and the inverse of the Mills ratio estimated from this equation was added as a covariate in the model for log-earnings. The result is model H_7, which produces almost no change in the inferences. Finally, H_8 incorporates the same selection-adjustment factor in the model including the new labor force scales (compare H_8 with H_5 in this instance). Although the selection covariate is significant in both regressions, there is virtually no change in the magnitudes of the coefficients for any of the new scales. In this case, there appears to be little value in adding explicit adjustments for sample selection (i.e., selecting only those respondents with earnings), and we conclude that our new scales reflect something quite different from sample-selection adjustments that might have been defined without reference to the labor force measures given here.

A logical difficulty with the models considered thus far is that they include effects of *current* labor force position when we are predicting earnings reported over the *previous* year. There is no simple way to sidestep the issue of incorrect time lags of this sort with the cross-sectional data used here. Current position in the labor force is highly associated with past year's labor market experiences, so much so that one should ask whether the current position measures need to be included once the prior experience measures have been included. In the 1982 CPS sample where these scales were derived, the first dimensions of current position and prior experience were highly correlated ($r = 0.856$) and the second dimensions were correlated moderately ($r = 0.570$).[13]

Deleting the current position measures produces model H_9 in Table 6.4, and we find that this model gives virtually the same conclusions as either one of the preferred models considered thus far (H_5 and H_8) and has virtually the same explanatory power. In our preferred model predicting occupational status (H_4 in Table 6.1), neither current position measure had a significant effect. Even though both effects are significant in a strict sense when log-earnings is considered, we would recommend deleting the current position measures *when modeling attainments from cross-sectional data*. Note that model H_9 still adds nearly 15% increment to the proportion of variance explained, nearly a doubling of the R^2 value of the standard Featherman–Hauser model (H_6). In other words, it is not necessary to include the current position measures in the occupation regressions (the effects were insignificant) or in the earnings regressions.

DISCUSSION

Our results suggest the importance of examining how labor force careers interact with and help to determine socioeconomic outcomes over the life course. This task has not been given high priority in contemporary research on the sociology of labor markets (see Kalleberg & Berg, 1987), and we hope that our results will encourage a reorientation of the research agenda. For regressions predicting occupational status, we found that the two measures of labor force experience increased explanatory power modestly and had strong effects in the predicted direction. The measures of current labor force position were not important once the other effects were included, which implies, for example, that there is no need to consider covariate adjustments for the occupational status of, say, part-time workers compared with full-time workers with the same occupational level.

The most compelling results, however, arose when considering log-earnings models adding the new scales. The increment in explanatory power is impressive with earnings regressions (a 15% increment to R^2). Almost all of the added effect could be attributed to the two measures of labor force experiences over the previous year. On the log scale, for example, a unit increase in our first prior experience measure is associated with the same magnitude of change in earnings as a 5.7-year increase in years of schooling (model H_9 in Table 6.2); the effects of both prior experience measures were the most important effects in the regression as judged by any of the standard criteria for relative importance. We believe we have made the case for seriously considering labor force experiences—or labor force histories of individual workers—as a major factor in the process that determines earnings.

Longitudinal analyses using measures of labor force experiences similar to those used here are the next step in this line of research. Important questions that need to be answered include these: (1) *Can the analogue to our prior experience measures be defined for longitudinal analyses?* We have implicitly assumed that the same scale should apply regardless of the worker's age, but perhaps some variability with age or time since completion of formal schooling ought to be allowed. If we attempt to use information from all or most items that capture the complexity of the labor force career, some type of scaling method will be needed. (2) *How might a model allowing for influences of the labor force career be specified?* One strategy is simply to define prior experience for each year in a cohort's experience and then create the scale summaries of

each of those. The most straightforward model of attainment level at year t for
a real cohort would then include the appropriate number of lagged prior experience variables, however scaled. Researchers could also define prior experience for the entire period between some starting point and the current age
of the worker. (3) *How can we model effects of labor force shocks (recessions or
upturns)?* Underemployment changes with recession and growth, which is one
of the main reasons why these measures can be called labor force measures. It
is possible that business cycle influences could be taken into account by either appropriate lagging of effects or possibly by letting the scale summaries
vary across periods of growth and recession.[14] An analysis like this would
allow us to chart whether labor force entry varies during periods of recession
and growth. (4) *Finally, how should we specify the causal chain of influences that
determine not only attainments at some point in time but also labor force career lines
throughout the life course?* The dramatically high level of flow among labor-
force states over the short term (Clogg, Eliason, & Wahl, 1990) must itself be
explained as part of our research agenda. These questions represent to us the
main challenges that lie ahead in bringing labor force activity into our attempts to understand social stratification.

NOTES

1. A more theoretical rationale for including the labor force in attainment models can be
found in Sørensen (1983). This important reference notes that something more is required than
simply adding a few dummy variables for labor force states like unemployed versus employed or
stable versus unstable labor force history. Schervish (1983) takes some labor force states (such
as types of unemployment) seriously as stratification-outcome variables in their own right, but
he does not link unemployment to occupational attainment, in spite of the fact that labor force
behavior of this kind is intimately connected to the process of occupational attainment. Spilerman
(1977) provides one approach for making this link, but he works with a limited set of measures
for labor force concepts (unemployed and employed) in an analysis of occupational careers and
career lines. Conventional models of status attainment typically ignore labor-force behavior
(e.g., see Blau & Duncan, 1967; Duncan et al., 1972; Hauser & Featherman, 1977; Featherman &
Hauser, 1978; Hauser et al., 1983; Jencks et al., 1983). Current models that specify influences of
economic sector, authority structures, and firm-specific characteristics can also be criticized on
the same grounds. In one important survey of labor market sociology produced two decades
ago (Kalleberg & Sørensen, 1979), labor-force concepts and measures like the ones we have in
mind are conspicuous by their absence. A more recent survey, which includes an integration of
much labor market analysis conducted during the 1970s and 1980s, excludes labor force concepts almost entirely (Kalleberg & Berg, 1987, see especially p. xii).

2. The scaling operation produces metric scores that explicitly adjust for sample selection
(Heckman, 1979), because even those who are outside the labor force are included in the table,
or model, used to estimate scores.

3. Note that both models compare favorably with conventional models of occupational
attainment reported, for example, in Featherman and Hauser (1978) in terms of R^2, in spite of
the fact that background variables have not been included.

4. However, an alternative interpretation would be that current position scores represent a contemporaneous "covariate" that should be included solely to adjust measures of occupational status (Duncan SEI) for current labor force position. This view would be appropriate, for example, if we would assign more status to full-time workers compared with part-time workers in the same occupation. Taking this view, we could regard the residuals from the regression of the SEI on both Current position scores as a reflection of "true" occupational status "net" of labor force position. By taking into account labor-force activity that amounts to less than full-time/year-round work, there seems to be little difficulty defending the ordering of causal factors here.

5. The effective range of these scores for those who were economically active over the previous year is about 2, so contrasts between the extreme scores therefore result in a difference of about 5.6 points on the SEI.

6. This can be contrasted with other techniques for doing this (see Stevens & Hoisington, 1987).

7. An interesting question is the degree to which prior experience and current position scores "share" variance with the other predictors included in the models. As noted above, adding these scores increases R^2 by 2.4% for men and by 3.1% for women. This might appear modest, but the effects of these variables have nevertheless been shown to be of dramatic importance when compared with other effects in the model. Both sets of scores are best viewed as variables that intervene between the other exogenous variables included in the models and current occupational status, and these scores are indeed associated with those other variables. It is natural to examine alternative partitions of explained variance. When regressions are estimated with just the prior experience and current position scores (excluding the other predictors), the R^2 values are 9.8% for men and 7.0% for women. The so-called independent effect of these measures should, therefore, be between 2.4 and 9.8% for males and between 3.1 and 7% for women (see Kruskal 1987). These figures serve as a reasonable estimate of the range of variability in occupational attainment that can be attributed to variability in labor force behavior (as measured by the prior experience and current position scores). An elaboration of the causal chain exploiting the essential recursiveness of the set of variables included in these models would be required to decompose explained variance in a more satisfactory manner. The difference between these range limits, about 7.4% for men and about 3.9% for women, does underscore the point that a substantial fraction of the explained variance is actually shared between the set of exogenous variables included in the standard model and the set of scores that have been added to the standard model.

8. Also, the estimates reflect conventional adjustments for sample weights.

9. This is the usual proxy for "experience"; but this measures the gross amount of potential experience in the labor force and is hence very different from the measures of labor market experiences captured in our El and E2 measures.

10. Slight differences between these two models and those reported by Featherman and Hauser (1978, p. 235) arose, perhaps because of differences in the use of sample weights

11. Note that El has about 60% of the variability in the OCG-II sample (measured in terms of standard deviations) compared with the original, 1982 CPS file where the scale was developed. The restriction here to males aged 25–64 accounts for the reduction in variability.

12. If we assess "strength of effects" or "relative importance" with t ratios or standardized regression coefficients, E1 and E2 appear to be much less important than respondent's education and DUNCF but almost as important as any background variable except BLACK.

13. See Clogg, Eliason, and Wahl (1990, Tables 6, 7, and 12) for further evidence on this point, including disaggregation by age and sex and results for nine CPS files covering the interval from 1976 to 1986.

14. In the study by Clogg et al. (1990), however, it was shown that marginal distributions rather than scale values produced from association models changed the most over time.

APPENDIX 6.1. A BRIEF NOTE ON CAUSAL ORDER

The simplest necessary condition for causal analysis is expressed in the proposition that causes must precede effects in a temporal sense. The temporal referents for prior experience and current position are thus important to recognize, inasmuch as these justify the regression strategy we use. To explain current (i.e., time point t) levels of occupation, it is clear that prior experience can serve as a possible causal factor, because it refers to a summary of behavior over the year preceding t. It is undoubtedly the case, however, that occupation levels prior to t influence prior experience, but this problem cannot be resolved without panel data of some kind, and this possibility need not concern us if we restrict attention to the prediction of current levels of occupation.

The current position variable is measured here contemporaneously with occupation, that is, in the survey week in March 1982. The exception, noted above, is the income (1) category, which by itself qualifies as a t^*, rather than a time-point t, measurement. We regard the current position measure, as well as scored variables based on it, merely as covariates, rather than as explanatory or potentially causal factors. To some extent at least, this difficulty is just conceptual assuming that it makes sense to make an adjustment for current labor force positions in explaining occupational attainment. For example, including current position as a predictor means that we are essentially analyzing residuals from current occupation regressed on current position. With panel data, of course, such problems could be overcome.

7

Market Experiences and Labor-Force Outcomes:
Fifteen Years of Race and Gender Inequality, 1982 to 1996

INTRODUCTION

So far we have examined trends in labor force activity in the 1970s and 1980s. We have attempted to highlight the various ways of looking at labor force activity and the insights that this yields about the workings of the social stratification system in the United States. In this chapter we update this picture further into the 1980s and 1990s and examine trends in gender and racial inequality in labor market outcomes.

One of the most puzzling questions facing students of social stratification is the persistent labor market inequalities encountered by women and minorities. The optimistic predictions regarding the opening of the social stratification system in the 1960s and 1970s (see Blau & Duncan, 1967; Duncan et al., 1972; Featherman & Hauser, 1978) have been called into question by research that points to persistent discrepancies between the labor market positions of white men compared with other groups. Persistent race and gender inequality in the labor market has drawn substantial attention from scholars attempting to explain its persistent nature (Reskin & Roos, 1987; Wilson, 1987, 1997; Petersen & Morgan, 1995; Holzer, 1996; Ridgeway, 1997; Hauan, Landale, & Leicht, 2000).

Much of this recent work has been complicated by upheavals in the labor market caused by the recession of the early 1980s, deindustrialization, and the subsequent reorganization of the workplace in the 1980s and early 1990s. Ethnographic and community-based accounts of these upheavals point to rapidly changing economic fortunes and life chances for manufacturing workers, growing legions of service sector workers, persistent and chronic unemployment for vulnerable groups, and general difficulties in gaining access to good jobs for new entrants to the labor force (Massey, 1990, 1996; Kasarda, 1991; Carnoy, 1994; Gordon, 1996; *New York Times*, 1996; Wilson, 1996; Phelps, 1997).

As part of this research effort, scholars have shown in numerous ways that labor market experiences (such as prior spells of unemployment, part-time and full-time employment patterns, and search behavior) have important consequences for labor force outcomes (subsequent employment, earnings, hiring decisions, working hours, and decisions to leave the labor force).[1] Research also suggests that the growing instability of labor markets has reduced steady employment opportunities for significant segments of the labor force. Still others suggest that once-steady regularities in the relationship between prior labor market experience and labor-force outcomes have been severed (Gordon, 1996; *New York Times*, 1996; Phelps, 1997; Leicht, 1998).

In this chapter we provide an empirical analysis of the association between annual labor market experiences and labor force outcomes using the 1982 to 1996 March Current Population Surveys. We draw on some of the methods used to produce the results we have discussed so far, particularly the purging methods discussed in Chapter 3. We also draw on RC association models and related methods developed by Clogg (1982d), Goodman (1986, 1991), Becker and Clogg (1989), Clogg, Eliason, and Wahl (1990), and Clogg and Eliason (1990) and discussed in Chapter 5. The analysis in this chapter yields some interesting, policy-relevant findings that contribute to our understanding of the workings of labor markets in the 1980s and 1990s. Specifically, we find that:

1. The relationship between prior labor market experiences and labor-force outcomes is remarkably stable from 1982 to 1996 and does not vary substantially by race or gender.
2. While the *relationship* between prior labor market experiences and current labor force outcomes appears stable, race and gender inequality in prior labor market experiences lead to clear differences in labor-force outcomes. This produces a temporally stable labor market segmented by race and gender.
3. Nonblack men dominate full-time/year-round employment.
4. Nonblack women dominate year-round voluntary part-time employment.[2]
5. Black men dominate unemployment combined with long, unsuccessful job searches.

6. Black women dominate another pocket of unemployment characterized by shorter unsuccessful job searches.

Using the purging methods developed by Clogg and Eliason (1988) and
Clogg, Shockey, and Eliason (1990), we examine the extent to which race and
gender inequalities in labor-force outcomes can be reduced by equalizing
market experiences across groups. We give our counterfactual scenario a concrete referent by standardizing labor market experiences of all groups on the
experiences of white men in 1996. Our results show that, had nonblack women
and black men and women the same labor market experiences as nonblack
men in 1996, *race–sex inequalities in labor force outcomes would have been reduced to 69% of observed inequalities.* This translates into over 16 million more
nonblack women, over 800,000 more black men and almost 2 million more
black women adequately employed full time.

MARKET EXPERIENCES AND INEQUALITY
IN LABOR FORCE OUTCOMES

To date a substantial amount of research has shown how market experiences play an important role in an individual's ability to find a job. Research
findings further suggest that persistent differences across groups in labor force
outcomes such as wages and employment status are due in large part to group
differences in labor-market experiences. Some also suggest that employers,
as well as the wider structure of labor markets, differentially reward the market experiences of different groups of workers, thus creating observed group
differences in the *relationship* between labor-market experiences and labor
force outcomes.[3]

Using Current Population Survey data, Clogg, Eliason, and Wahl (1990)
showed how, at the national level, market experiences affect labor force outcomes. This work was discussed in Chapter 5. They found that in 1982, the
experience/outcome relation could be segmented into distinct pockets of association. Those who were not working and not looking for work throughout
the year were highly likely to be out of the labor force a year later. Those who
had year-round full-time employment were highly likely to be adequately full-
time employed a year later. Those who were not working but looking for work
off and on throughout an entire year were likely to be an unemployed new
entrant or reentrant to the labor market a year later. And those who were
voluntarily part-time employed throughout the year were highly likely to be
so a year later.

In case studies of Chicago-based organizations, Bills (1988) provided a
closer look at the way in which market experiences are important in obtaining jobs. His work showed that managers put far more emphasis on experi-

ence than on educational credentials when hiring for nonmanagement positions (Bills 1988). Interviews, test scores, and personality all played more important roles than did educational credentials for these positions.

Blacks are particularly disadvantaged in the hiring process, in part because of employers' perceptions of their market experience. Holzer's work (1996) shows, for example, that employers are less trusting of those with unstable work histories and more trusting of those with more credentials. Blacks are disadvantaged in part because employers perceive that they have less stable work histories and fewer credentials. This general perception appears to hold for women (compared with nonblack men) as well. Combining these two tendencies, then, black women are doubly disadvantaged in gaining the trust and confidence of employers (see Reskin & Roos, 1990).

Wilson's research (1987, 1996) further suggests that blacks in inner cities are particularly disadvantaged in terms of employment, thus hindering their ability to gain valuable job experience and to develop stable work histories. More precisely, Wilson (1996) shows a growing tendency for inner-city black males with low skill levels to have a difficult time finding adequate employment. Wilson finds that employers will label blacks in general as "lazy" (Wilson, 1996, p. 112) and "ignorant" (Wilson, 1996, p. 122), and often have a hard time hiring blacks because of their poor background, poor education, poor attitude, and/or poor self-presentation. Employers also express fear and suspicion as reasons for not hiring blacks.

These general perceptions are unfortunately not without reinforcing empirical evidence. Using Census data from 1890 through 1990, Vedder and Gallaway (1992) show that the white–nonwhite unemployment ratio has not always been as high as it currently is. Prior to World War II (back through 1890), that ratio in the United States barely moved from the null value of 1. After World War II, however, the ratio was often around 2, and never below 1.5. Farley's (1984) work further shows that from 1960 to 1982, labor force nonparticipation rates and unemployment rates for blacks were considerably higher than those for whites.

These results show that, after World War II, blacks were more likely to have experienced unemployment and nonemployment than were nonblacks. However, employers' reluctance to give individual blacks a chance based on general perceptions feeds into unstable early work histories, which in turn reinforces employers' already held stereotypes of blacks. It is in this regard that blacks' market experiences and employers' perceptions tend to be mutually reinforcing.

Although employer motivation is different, the social closure argument presented by Tomaskovic-Devey (1993) results in similar employer actions. Here, rather than fear and suspicion as forces driving employers to shun blacks in the employment arena, Tomaskovic-Devey argues that employers in part seek to protect their existing privilege. This class-based fear is not the same as the employer fears described in Wilson's research. But the outcome is still the same. Blacks end up locked out of adequate full-time employment relative to other groups, especially white men, and are subsequently hindered from developing a history of stable full-time employment.

Work by England (1992) and England and Farkas (1986) highlight that employers also have reservations about women's commitment to the labor market. This argument is put forward by others as well (e.g., Reskin & Roos, 1990). These researchers also suggest that employers reserve certain jobs for men and others for women, based largely on the gender of the tasks necessary to be successful in the job.

Though developed mainly with reference to gender segregation, employers' orientation toward both blacks and women in the market can be well explained by Ridgeway's theory of categorization (1997), an idea not completely unlike notions of statistical discrimination. Although it is beyond the scope of our chapter to elaborate the theory here, these ideas provide a strong conceptual tool for understanding, and predicting, discriminating behavior. In an oversimplification of Ridgeway (1997), sex is often a noticeable attribute that individuals use to categorize self and others in interactions. Because sex is a dichotomous attribute (one can easily be classified as "man" or "woman"), it is one of the primary attributes used to draw inferences. In the United States, the black–white race distinction embedded in our country's history is likely to render race as a primary attribute as well. From this reasoning, employers will often use these primary categorizations such as race and sex to infer other properties of an individual. Because gathering complete information on each potential employee is costly, both in terms of money and effort, this inference on the part of employers reduces long-term costs.

Although we cannot address the full richness of these arguments, much of this work points to employers' differential treatment of market experiences across race and gender groups. More precisely, this work suggests that we should expect the relationship between labor market experiences and labor force outcomes to be segmented along race and gender boundaries. We should see tighter connections between prior work histories and current labor force positions for women and blacks compared with nonblack men. This is because employers should have the tendency to require stronger and more stable work histories of women and blacks to allow them entry into adequate full-time employment. This tendency should appear as a stronger positive association between prior stable work histories and subsequent adequate full-time employment for women and blacks. Because those with unstable work histories are locked out of adequate employment, this same process should result in a stronger positive association between having an unstable work history and subsequent unemployment for these groups as well.

To examine the association between market experiences and labor force outcomes for black and white men and women, we use data from the 1982 through the 1996 March Current Population Surveys (CPS). We use the market experience measure discussed in Chapter 5, and derived from the work of Clogg, Eliason, and Wahl (1990). This is a 16-fold classification measuring the experiences that respondents had had over the course of the year prior to each March CPS survey. Our labor force outcome measure is the expanded LUF measure discussed in Chapter 5 as well, and is that used by Clogg, Eliason, and Wahl (1990). Note that the only difference between our measure here and that used in Clogg, Eliason, and Wahl (1990) is that the educational mismatch category has been dropped from the analysis. This was done because the occupational classifications, on which the mismatch measure is based, change during the time period between the 1982 and 1996 Current Population Surveys.

Appendix 7.1 discusses details of the models and methods used to derive the results we discuss in this chapter. In the next section we present the distributions of the observed labor force outcomes from 1982 to 1996. We then turn our attention to the relationship between market experience and labor force outcomes. We then present results that adjust the labor market experiences of blacks and women to reflect the past labor market experiences of white men.

Observed Trends in Labor Force Outcomes

Figures 7.1 through 7.9 give the trends in labor force outcomes for the four race–gender groups. These figures show the odds of each labor force outcome for each race–gender group from 1982 to 1996. Figure 7.1 presents

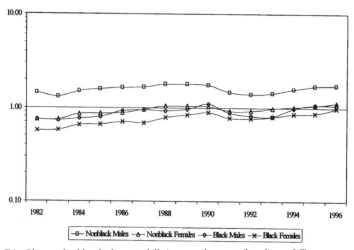

Figure 7.1. Observed odds of adequate full-time employment (baseline = 9.7).

the odds of adequate full-time employment for each race and gender group. The solid line in the middle of each graph (normed to 1 in each graph for comparison purposes) represents the average baseline odds. Lines for each race–gender group represent deviations around the baseline value. Odds above the baseline suggest that the group has a greater than average chance of, in this case, adequate full-time employment. Odds below the baseline suggest that a group has a less than average chance of adequate full-time employment. This interpretation also applies to the other labor force outcomes we examine in Figs. 7.2 through 7.9 as well.

While we norm the baseline odds in the graphs to 1, the reader should understand that this line represents some average value. In Fig. 7.1, this baseline represents the value 9.7. That is, the average odds of adequate full-time employment, relative to all other labor-force positions, is 9.7 over this time period. By comparing these baseline odds to other baseline odds, we obtain a measure of the average odds of being adequate full-time employed relative to some other labor force position. For example, the baseline odds for low-income full-time employment is 1.3. Thus, without regard to any race–sex classification, the average individual in the sample is [9.7/ 1.3=] 7.5 times more likely to have adequate versus low-income full-time employment.

The distribution of the odds of adequate full-time employment over time presented in Fig. 7.1 separates nonblack men from all other race–gender groups. On the whole, nonblack men are 1.5 times more likely to have adequate full-time employment than the other race–gender groups. For all groups, however, there is a slight trend upward, except during the years 1989 to 1992.

Figure 7.2 suggests that, for low-income full-time employment, the four

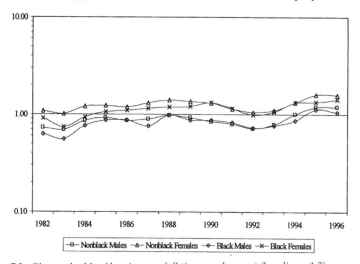

Figure 7.2. Observed odds of low-income full-time employment (baseline = 1.3).

race–gender groups do not separate very much. If anything, there is a slight gender split, with women having slightly higher odds than men. The exception to this is in the early 1980s where the odds for black females are closer to the male odds. With the exception of the 1989–1992 period, there is a slight trend upward in the odds of low-income full-time employment.

Figure 7.3 suggests that voluntary part-time employment is dominated by nonblack women and is least likely among black men. There is very little overall trend in this picture, but the trend line for nonblack men and black women converges around the average odds after 1990. These results suggest that voluntary part-time employment is linked to relative economic prosperity and traditional gender roles.

Figure 7.4 shows the odds of having part-time employment for economic reasons (e.g., slack work). Here it is difficult to see any consistent distinction among groups or much of any trend over time. In 1982, black males have the highest odds of having this type of part-time employment, with nonblack females having the lowest odds. By 1986, black males still have the highest odds, but the other groups have coalesced. After 1989, though, nonblack and black males tend to trend together, creating a slight gender effect (except for 1992). However, by 1996 nonblack males are now above the baseline odds, with the other three groups below.

Figure 7.5 suggests that there were significant changes in the relationship between the race and gender groups involved in involuntary part-time work. The 1982 recession appeared to increase the odds of involuntary part-

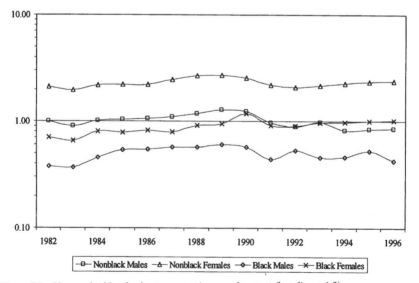

Figure 7.3. Observed odds of voluntary part-time employment (baseline = 1.3).

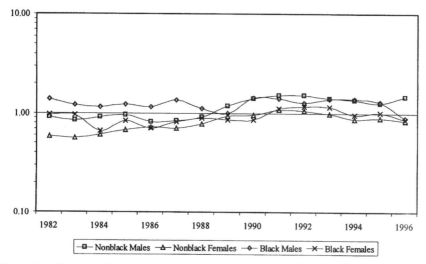

Figure 7.4. Observed odds of part-time employment due to economic reasons (baseline = 0.4).

time work for all groups, especially black women. By 1990, all groups except black women were below the baseline odds on this measure. There is a rise with the early 1990s recession and a return to the late 1980s pattern. Overall these results suggest that black women are most likely to be involuntarily part-time employed because of not finding full-time work.

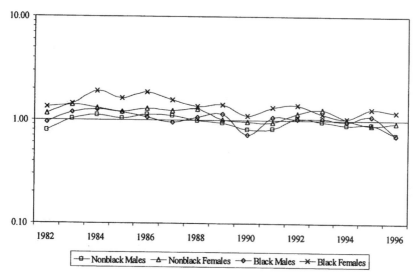

Figure 7.5. Observed odds of part-time employment because no full-time work is available (Baseline = 0.4).

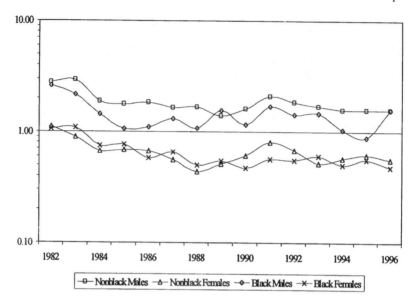

Figure 7.6. Observed odds of unemployment due to layoffs (baseline = 0.2).

Figure 7.6 clearly shows, in the 1982 recession, a gender gap in the odds of being laid off, with males hit harder than females. By 1985, this gap dissolves temporarily, as black men's odds decline close to female levels, while nonblack men's odds remain roughly as they were in 1984. By 1987, the gender gap reappears, shifts a bit in 1988, but rebounds yet again in 1989. With the exception of 1995, this persistent gender gap is apparent from 1989 to 1996.

Figure 7.7 presents the odds of unemployment due to quits and losses. Black men have the highest odds throughout this period, and nonblack women have the lowest odds. Again, the aftermath of the 1982 recession appears to have created the most hardship, with black men's odds rising above 2 times the baseline in 1983. Coupled with the 1982 recession, the Bush Presidency appears to have created the widest disparity here. By 1995, however, the gap between these groups closes considerably, but then appears to widen again in 1996. The overall picture of unemployment is consistent with most of the prior work on unemployment in the 1980s and 1990s; there is a strong and persistent gap in unemployment between black men and all other groups that only subsides in the mid 1990s. Nonblack women are least likely to quit or lose their jobs throughout the period.

There is a persistent race gap in new entrant and reentrant unemployment displayed in Fig. 7.8. By contrast, there is a persistent gender gap in labor force inactivity displayed in Fig. 7.9. Women are consistently more likely

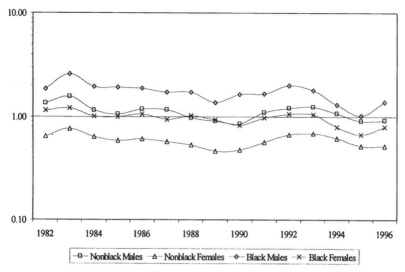

Figure 7.7. Observed odds of unemployment due to quits and job losses (baseline = 0.6).

to be out of the labor force than men, regardless of race, and there is little in the way of a race gap by the end of the period. There is a significant gap between black and nonblack women at the beginning of the 1980s. This initial gap and its subsequent disappearance may suggest that voluntary withdrawal from the labor force was the prerogative of nonblack women. As earn-

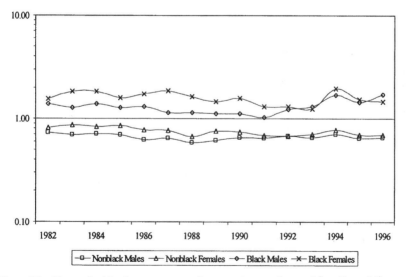

Figure 7.8. Observed odds of new entrant and reentrant unemployment (baseline = 8.1).

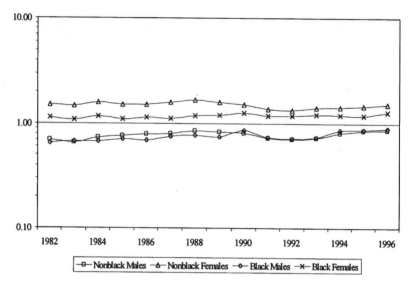

Figure 7.9. Observed odds of not being in the labor force (baseline = 8.1).

ings stagnated during the 1980s the labor force participation gap between white and black women converged as dual earner households moved from nominally acceptable to necessary for the maintenance of a middle-class lifestyle (see Schor, 1993).

Decomposing the Experience-Outcome Association

In this section we examine whether the linkages between prior labor market experience and current labor force positions are different across the four race–gender groups. One issue that is fundamental to prior labor market research is the question of labor market segmentation. Does the labor market produce systematically different experiences for a diverse work force, experiences that are regularized and repeatable? This section examines whether there are differences by race and gender in the ways that prior labor force experience translates into current job opportunities.

Here we present results from the decomposition of the association models discussed in Appendix 7.1. Details of these results are presented in Appendix 7.2. We decompose the association between prior experience and current labor force position into (1) a *group-specific component* that takes into account the distinctive relationships between labor force experience and outcomes for different race and gender groups; and (2) a *time-specific component* that takes into account changes in the experience/outcome relationship over time from 1982 to 1996.[4]

These decompositions suggest that the relationship between prior labor market experiences and current labor force outcomes is remarkably homogeneous and stable. Less than 1% of the total association between market experiences and labor force outcomes can be accounted for by race–sex differences in the association. Similarly, temporal changes account for less than 1% of the total association as well. We arrive at these figures in the following way.

All of our calculations start with the conditional independence model, which forces the relationship between prior experience and current outcomes to be independent of one another for each race–gender group and for each time period. This model is of no particular analytical value in itself, but it does provide a baseline likelihood-ratio statistic that can be used to assess the other substantive models. This likelihood ratio statistic is large (1,990,852) because the "no-relationship-between-experience-and-outcomes" assumption does not describe the data well. The relative improvement gained in our substantively interesting models can be assessed by comparing the likelihood ratio values of these subsequent models to the baseline given by the conditional independence model.

Our first analysis estimates a model that allows for one and only one relationship between prior labor force experience and current labor force positions. This is similar to assuming that there is only one labor market for all groups and for all time periods. The likelihood ratio statistic for the one-dimensional solution for this model is 300,265. Although this model provides a poor fit by classical standards, it does account for a considerable 84.92% of the total association in the table.[5] A second dimension accounts for an additional 9.23% of the association, while a third dimension accounts for an additional 4.38%. Once we move beyond three dimensions, however, the proportional contribution drops dramatically, with less than 1% accounted for by dimensions four through eight for these complete homogeneity models.

The proportion left over from this homogeneity model is the proportion of the total association due to differences across race-sex groups and shifts in the association over time. The proportion of the total association that is due to the heterogeneity across race-gender groups and over time combined is 1.09%, with 0.77% due to differences across race-gender groups and 0.80% due to differences over time.[6] Thus, our results show that the association between market experiences and labor force outcomes appears extremely stable over time and across race–gender groups.

Table 7.1 presents the influence measures that describe the relationship between specific prior experiences and current labor-force positions. The specific method used to derive these measures is presented in Appendix 7.3. Using the guideline discussed in Appendix 7.3, those influence measures that are greater than two times the standard deviation in absolute value,

Table 7.1
Influence Measures from the Race, Gender, and Temporal Homogeneity Model [RC(3)].

| | Current labor force position | | | | | | | | |
| | Unemployed | | | | Part time | | | Full time | |
Labor market experience	NILF	New/ reentrants	Quits/ losses	Layoffs	No full time	Economic reasons	Voluntary	Low income	Adequate
Nonworkers									
Not looking	**3.77**	1.07	-0.59	-0.90	-1.47	-1.35	0.45	-0.52	-0.46
Looked 15+ weeks	0.44	1.29	1.25	1.01	-0.12	-0.15	**-1.91**	-0.95	-0.85
Looked 1-14 weeks	1.57	1.13	0.23	-0.02	-0.26	-0.48	-0.38	-0.65	-1.13
Part year–Part time									
Not looking	1.19	-0.03	-1.10	-1.15	0.07	-0.29	1.75	0.27	-0.71
Looked 15 + weeks in 1 stretch	-0.51	0.27	0.03	-0.01	0.78	0.32	0.27	-0.09	-1.06
Looked 15+ weeks in 2+ stretches	-0.52	0.40	0.27	0.21	0.70	0.30	-0.13	-0.22	-1.01
Looked 1-14 weeks in 1 stretch	-0.05	-0.08	-0.62	-0.61	0.61	0.16	1.20	0.23	-0.84
Looked 1-14 weeks in 2+ stretches	-0.18	0.09	-0.30	-0.31	0.60	0.19	0.70	0.06	-0.85
Part year–Full time									
Not looking	1.13	-0.28	-0.34	-0.32	-1.03	-0.55	0.13	0.15	1.11
Looked 15+ weeks in 1 stretch	-0.56	0.18	0.96	0.95	-0.29	0.07	-1.60	-0.33	0.62
Looked 15+ weeks in 2+ stretches	-0.65	0.32	1.00	0.97	-0.08	0.15	-1.58	-0.40	0.26
Looked 1-14 weeks in 1 stretch	-0.60	-0.27	0.56	0.64	-0.40	0.06	-1.04	-0.02	1.08
Looked 1-14 weeks in 2+ stretches	-0.72	-0.19	0.54	0.61	-0.17	0.15	-0.93	-0.04	0.76
Full year									
Voluntary part time	-0.23	-1.32	-1.67	-1.45	0.36	0.16	**2.67**	1.08	0.40
Involuntary part time	**-3.18**	-0.97	0.05	0.31	1.66	1.24	0.48	0.60	-0.18
Full time	-0.90	-1.61	-0.26	0.07	-0.94	0.02	-0.07	0.81	**2.86**

Note: Entries in bold are greater than two standard deviations from the mean. See Appendix 7.1 for details.

$\left|\hat{w}_{ij(kl)}\right| > 2s_{w_{(kl)}}$, are shown in bold in Table 7.1. Those experience/outcome combinations for which $\left|\hat{w}_{ij(kl)}\right| > 2s_{w_{(kl)}}$ have a relatively high contribution of moving the data away from independence.

What becomes clear from Table 7.1 is that there are only a few influence measures that stand out. Probably the most noticeable are the two ends of the main diagonal. The upper left entry of 3.77 indicates a strong positive relationship between nonworkers not looking for work over the previous year and those currently out of the labor force at the time of the survey. The lower right entry of 2.86 indicates that year-round full-time workers are highly likely to have adequate full-time employment at the time of the survey. The remaining strong positive entry, 2.67 in row 14, column 7, links year-round voluntary part-time workers with those voluntarily part-time employed at the time of the survey.

There are also two strong negative associations in the table. The first (–3.18 in row 15 column 1) indicates that those who are involuntarily part-time employed are unlikely to be out of the labor force the following year. The other strong negative association (–1.91 in row 2, column 7) indicates that those who were not working, but looking for work for 15 or more weeks were unlikely to be voluntarily part-time employed at the time of the survey.

While these results may not strike the reader as surprising, they do reveal a few aspects of the workings of labor markets that are worth pointing out:

1. Those "all the way in" and "all the way out" of the labor market are in very stable positions. Leaving the labor force does not appear to be a year-to-year transitory phenomenon.
2. Those who are employed part time on a voluntary basis seem to occupy a stable position in the labor market as well. Moving into and out of this labor force position seems to be rare; rather the occupants of this position tend to stay there from year to year.
3. Those who are involuntarily part-time employed tend to continue to look for full-time employment and tend not to leave the labor force as discouraged workers.
4. Those out of the labor force who put energy into long searches for full-time employment are unlikely to accept part-time employment of any kind.

Market Experiences of Race–Gender Groups

Our results suggest that the relationship between labor market experiences and labor-force outcomes is remarkably stable across race–gender groups and over time. Because of this stability, observed differences in labor force outcomes across race–gender groups over time are not due to differences in the experience–outcome relationship, as these differences scarcely exist. Instead, it is the differences in the market experiences themselves of blacks and nonblacks and men and women that determine corresponding differences in labor force outcomes.

Because the 16 labor-market-experience types produce complex trends from 1982 to 1996 that differ by race–gender groups, examination of the complete set of trends would be too complicated for this chapter.[7] However, a comparison of the first and last years in the series provides insight into race–sex differences in market experiences that tend to persist throughout this period, and that are consequential for the observed differences in labor-force outcomes.

Table 7.2 gives the baseline odds and the race–sex group odds for the 16 labor-market experiences for the 1982 and 1996 CPS data. Overall the odds of gaining year-round full-time employment has increased by 1.5 times (= 22.37/ 14.77), but this increase is not equally distributed across subgroups. Black men's share of this increase has decreased to 89% of their 1982 level and nonblack men's share has also decreased to 74% of their 1982 level. Black women's share has increased by 24%, while nonblack women's share has increased by 6%. Relative to the other race–gender groups, black women appear to be gaining greater access to year-round full-time employment. However, nonblack men by far have the highest odds of year-round full-time employment in both 1982 and 1996, as well as throughout the remaining years in this time frame.

The odds of an extended (15+ week) job search among nonworkers has barely changed from 1982 to 1996. Moreover, black men are considerably more likely than others to job search for long periods without working. In 1982 they were 1.6 times more likely than black women, 4.1 times more likely than nonblack men, and 5.4 times more likely than nonblack women to engage in long job searches without working. In 1996, those numbers are 1.6, 3.7, and 4.9 respectively. Black men are also far less likely to experience year-round voluntary part-time employment.

Nonblack women dominate the ranks of the year-round voluntarily part-time employed. Other likely market experiences for nonblack women include having part-year part-time employment while not looking, part-year part-time employment while looking 1–14 weeks in one stretch, and not working while not look to 14 weeks.

As indicated in Table 7.1, comparing the most likely labor-force outcomes to the market experiences of race–gender groups reveals the following race/ gender stratification in the market. Nonblack men dominate the year-round full-time and adequate employment niche (row 16, column 9 in Table 7.1). Nonblack women dominate the year-round voluntary part-time employment niche (row 14, column 7 in Table 7.1). Black men dominate the chronically

Table 7.2
Labor Market Experience Odds for 1982 and 1996, Baseline and by Group

Labor market experience	1982					1996				
	Baseline odds	Nonblack males	Nonblack females	Black males	Black females	Baseline odds	Nonblack males	Nonblack females	Black males	Black females
Nonworkers										
Not looking	12.56	0.68	1.59	0.79	1.17	15.62	0.78	1.37	0.82	1.14
Looked 15+ weeks	0.52	0.59	0.45	2.42	1.56	0.57	0.63	0.48	2.33	1.42
Looked 1–14 weeks	0.48	0.46	0.94	0.91	2.54	0.41	0.63	0.92	1.11	1.56
Part year–part time										
Not looking	2.32	0.96	1.97	0.63	0.84	2.77	0.93	1.79	0.66	0.91
Looked 15 + weeks in 1 stretch	0.38	0.82	0.95	1.29	0.99	0.40	0.87	0.96	1.11	1.09
Looked 15+ weeks in 2+ stretches	0.31	1.15	0.67	1.36	0.96	0.26	1.10	0.88	0.96	1.08
Looked 1–14 weeks in 1 stretch	0.41	0.92	1.58	0.58	1.18	0.34	0.96	1.74	0.65	0.92
Looked 1–14 weeks in 2+ stretches	0.26	0.87	1.11	1.18	0.88	0.21	1.04	1.07	0.95	0.95
Part year–full time										
Not looking	2.42	1.25	1.41	0.67	0.84	2.91	1.11	1.29	0.75	0.94
Looked 15+ weeks in 1 stretch	1.00	1.22	0.66	1.46	0.85	1.02	1.11	0.68	1.28	1.04
Looked 15+ weeks in 2+ stretches	0.69	1.56	0.54	1.55	0.76	0.41	1.48	0.51	1.86	0.72
Looked 1–14 weeks in 1 stretch	1.50	1.29	0.85	1.20	0.76	1.11	1.27	0.90	0.97	0.91
Looked 1–14 weeks in 2+ stretches	0.56	1.59	0.75	1.26	0.66	0.37	1.38	0.67	1.46	0.75
Full year										
Voluntary part time	1.30	1.12	2.27	0.48	0.83	2.12	0.96	2.17	0.50	0.97
Involuntary part time	0.41	0.82	1.09	0.65	1.74	0.56	0.84	1.15	0.92	1.13
Full time	14.77	1.70	0.85	1.05	0.66	22.37	1.26	0.90	0.93	0.82

unemployed niche combined with long job searches (row 2, columns 2 and 3 in Table 7.1). Black women dominate the shorter-term unemployed niche (row 3, column 1 in Table 7.1). Thus, the US labor market between 1982 and 1996 appears to be distinctly and stably partitioned along these race–gender lines.

EQUALIZING MARKET EXPERIENCES ACROSS RACE–GENDER GROUPS

In this section we provide some insights into how labor-force outcomes would be distributed across groups had market experiences been equalized across race–gender groups. To obtain these counterfactual distributions, we use purging methods developed by Clogg and his associates (Clogg, 1978; Clogg & Eliason, 1988), and discussed earlier in the book. Here we use marginal CG purging as described in Clogg and Eliason (1988), treating market experiences as the composition variable and race–sex classifications as the grouping variable.

In obtaining the purged distributions, we use the market experiences of 1996 nonblack men as the standard, one of the more favorable market experience distributions over this time period.[8] For nonblack females, this would result in a shift away from part-time employment and nonemployment toward year-round full-time employment. More black females would experience year-round full-time employment than is currently the case as well, and fewer would experience periods of unsuccessful search. Similarly, fewer black males would experience extended periods of unsuccessful job searching. This would be accompanied by an increase in black males experiencing year-round full-time employment.

Marginal CG purging is used because it allows us to obtain labor force outcome distributions that would have been obtained had market experiences been, in an experimental design sense, orthogonal to race–sex groups. That is, we are able to answer the question of what the labor force distributions would have been had individuals been randomly assigned to market experiences, thus mimicking an experimental design setting. Comparisons of the marginal CG purged labor force outcome distributions with the observed outcome distributions allow us to assess the impact that the different market experiences of blacks and nonblacks and men and women have on labor force outcomes.

To obtain a precise measure of the overall race–gender inequality in the purged and observed labor force outcome distributions, we use a measure similar to that discussed in Goodman (1991) and used in Hout, Brooks, and Manza (1995). Though used by Hout et al. (1995) in assessing the relation between social class and voting, this measure, called kappa (κ), is useful in our case as well. Here this measure, calculated as the standard deviation of race–gender group differences in market outcomes, gives a summary of inequality in market outcomes across race–gender groups over time. Details on this measure of inequality are presented in Appendix 7.5.

Comparing Observed and Purged Distributions

Figure 7.10 presents the inequality measures for the observed and purged distributions. It is clear that equalizing market experiences across race–gender groups (as defined by marginal CG purging) decreases overall inequality in labor force outcomes. Using the overall measures (given at the bottom of Fig. 7.10), equalizing market experiences reduces inequality to about 69% of the observed. What is also apparent in Fig. 7.10 is that the reduction in inequality does not change all that much throughout the time frame. Inequality would be reduced by a maximum of 36% in 1984 and a minimum of 26% in 1996. We next examine similar distributions for each labor-force position.

Adequate Full-Time Employment. Figure 7.11 presents the odds of adequate full-time employment had market experiences been independent of race–gender groups. These purged odds are directly comparable to the observed odds given in Fig. 7.1. This comparison shows that the baseline has risen from an observed odds of 9.7 to a purged odds of 13.7. This indicates that, on average, the odds of adequate employment would have increased by

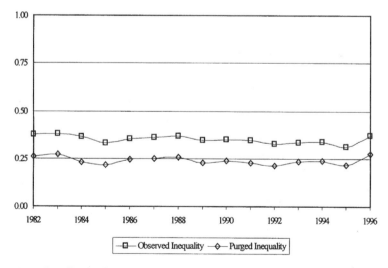

Figure 7.10. Overall inequality measures for observed and purged labor-force outcome distributions (overall observed inequality = 0.35; overall purged inequality = 0.24).

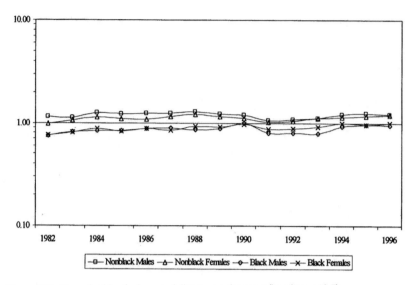

Figure 7.11. Purged odds of adequate full-time employment (baseline = 13.7).

a factor of [13.7/ 9.7=] 1.4 had market experiences been equalized across race–gender groups. Using population figures for 1996, the corresponding increase in adequately full-time employed is estimated to be over 16 million more nonblack women, over 800,000 more black men, and almost 2 million more black women in 1996.

A comparison of the observed and purged odds shows that the gap between the highest and lowest purged odds is smaller than the observed counterpart. This indicates that more equality in the odds of adequate full-time employment would be obtained had market experiences been equalized across race–gender groups. Figure 7.12 shows this more precisely by providing a comparison between the observed and purged inequality measures. Overall, inequality in adequate full-time employment would be reduced by 50% (= [0.14/ 0.28] × 100) had market experiences been equalized across race–gender groups. Although observed inequality declines over time, the purged inequality is always less than the observed.

Low Income Full-Time Employment. Figure 7.13 presents the purged odds for low-income full-time employment. This labor force category most closely approximates the idea of "working poverty" in prior research (see Hauan et al., 2000). The observed baseline odds for this market outcome is 1.3 (see Fig. 7.2). When compared with the purged baseline odds of 1.6, equalizing across-group market experiences actually increases the baseline odds slightly by a factor of [1.6/1.3=] 1.2. Ironically, the move toward equality in prior labor-

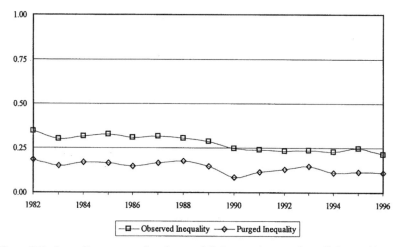

Figure 7.12. Inequality measures for adequate full-time employment (overall observed inequality = 0.28; overall purged inequality = 0.14).

market experiences would increase the odds of low-income full-time employment. We are not quite sure whether this is because more people would be employed full time who otherwise would not be or whether some incumbents in adequately compensated full-time positions would move to the "low-income" category.

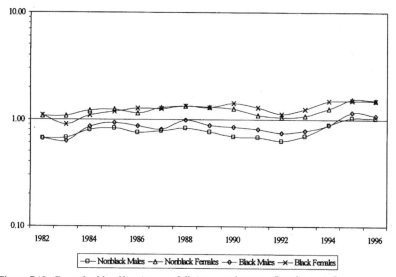

Figure 7.13. Purged odds of low-income full-time employment (baseline = 1.6).

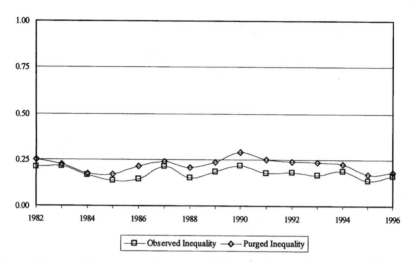

Figure 7.14. Inequality measures for low-income full-time employment (observed overall inequality = 0.18; overall purged inequality = 0.22).

The inequality measures presented in Fig. 7.14 further indicate that equalizing across-group experiences would lead to an increase in inequality for low-income full-time employment. The overall inequality measures show that equalizing across-group experiences would increase inequality by a factor of 1.22, or 22%. Further, except for 1983 and 1984, purging leads to increased inequality throughout the rest of the time periods.

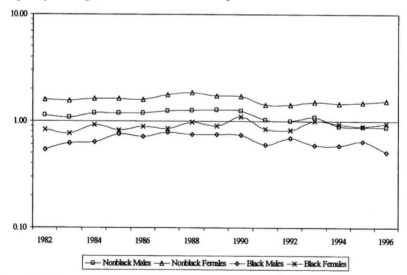

Figure 7.15. Purged odds of voluntary part-time employment (baseline = 1.2).

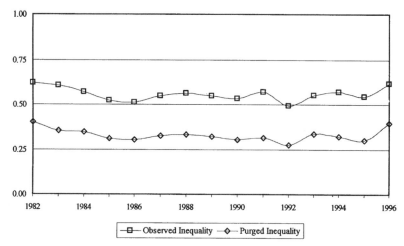

Figure 7.16. Inequality measures for voluntary part-time employment (overall observed inequality = 0.56; overall purged inequality = 0.33).

Voluntary Part-Time Employment.   Although the baseline odds do not change all that much between the observed and purged odds of voluntary part-time employment (1.3 and 1.2 in Figs. 7.3 and 7.15, respectively), Fig. 7.16 shows that inequality is greatly reduced in the purged odds. Overall inequality would have been reduced by 41% (= [1 − (0.33/0.56)] × 100) had market experiences been equalized across groups. The individual inequality measures also show that there appears to be a substantial reduction in inequality throughout the entire time period. This reduction in inequality is due in large part to a reduction in the odds that nonblack women would be voluntarily part-time employed had market experiences been equalized across race–gender groups.

Part-Time Employed Due to Economic Reasons.   Equalizing the across-group differences in market experiences does little to change either the baseline odds or inequality for part-time employment due to economic reasons. Both the observed and purged baseline odds are 0.4 (compare Figs. 7.4 and 7.17, respectively). The highest observed disparity is in the early 1980s between black men and nonblack women. Further, there is no difference in the overall observed and purged inequality measures. Examination of the individual inequality measures in Fig. 7.18 further reveals little difference in the across-time inequality for part-time employment due to economic reasons.

Perhaps of interest, though, is that in the early 1980s, during the Reagan presidency, the purged distributions show either more or relatively equal levels of inequality than the observed distributions, with 1983 revealing a noticeable gap. From 1989 onward, through the Bush and Clinton presidencies, this

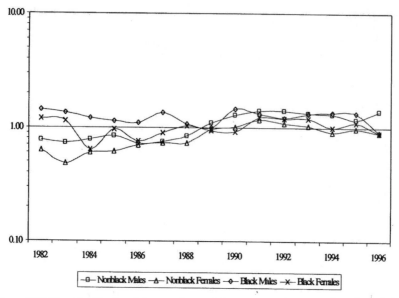

Figure 7.17. Purged odds of part-time employment due to economic reasons (baseline = 0.4).

tendency no longer holds. During that time period, orthogonalizing the relation between market experiences and race–gender groups results in a reduction in inequality for part-time employment due to economic reasons.

    Part-Time Employed Due to No Full Time Available.   As with part-time employed for economic reasons, here there is little difference in either the

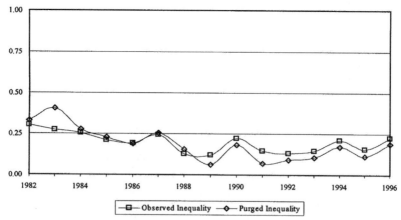

Figure 7.18. Inequality measures for part-time employment due to economic reasons (overall observed inequality = 0.21; overall purged inequality = 0.21).

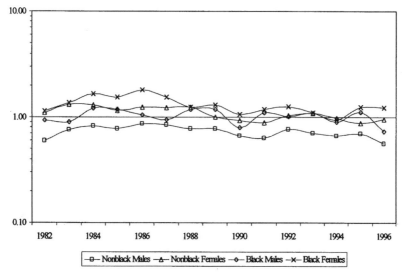

Figure 7.19. Pruged odds of part-time employment because no full-time work is available (baseline = 0.3).

baseline odds or the distances among group odds. The observed baseline is 0.4, while the purged baseline is 0.3 (compare Figs. 7.5 and 7.19, respectively). The overall inequality measures also differ very little, indicating a slight increase in inequality from 0.18 for the observed distribution to 0.23 for the purged. Figure 7.20 further shows that inequality would be slightly increased

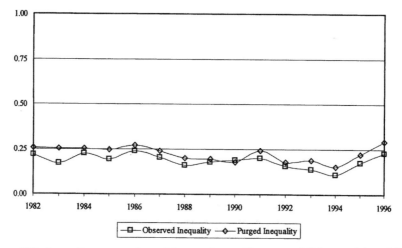

Figure 7.20. Inequality measures for part-time employment because no full-time work is available (overall observed inequality = 0.18; overall purged inequality = 0.23).

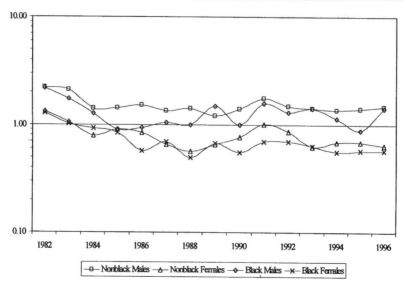

Figure 7.21. Purged odds of unemployment due to layoffs (baseline = 0.2).

for most of the 1982 to 1996 period had market experience been independent of race–gender groups.

Unemployed Due to Layoffs.   Once again, there is little difference between the purged and observed odds for unemployment due to layoffs (compare Figs. 7.6 and 7.21 respectively). There is no shift in the baseline odds, with both at 0.2. However, Fig. 7.22 shows that, had market experiences been independent of race–gender groups, overall race–gender inequality would be reduced 29% (= $[1 - (0.34/0.48)] \times 100$) for those unemployed due to layoffs. This reduction is also evident throughout the period. However, toward the end of the time frame, from 1993 to 1995, we can see that observed inequality begins to decline, closing the gap between the purged and observed inequalities. In 1996, though, this trend is reversed, as observed inequality shoots to its maximum, along with the purged inequality.

Unemployed Due to Quits and Losses.   For unemployment due to quits and losses, we see that, although the baseline does not change at all between observed and purged odds (compare Figs. 7.7 and 7.23, respectively), inequality does appear to be affected by orthogonalizing market experiences across race–gender groups. This reduction in inequality is evidenced by the compression of the purged odds toward the baseline relative to the observed odds, though black men are still more susceptible to unemployment due to quits

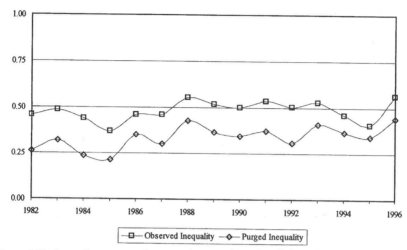

Figure 7.22. Inequality measures for unemployment due to layoffs (overall observed inequality = 0.48; overall purged inequality = 0.34).

and losses even in the purged case. The inequality measures in Fig. 7.24 further show that, had experiences been independent of race–gender, overall inequality would have declined 44% (= [1 − (0.22/0.39)] × 100) for this labor force outcome. An examination of the trend also reveals that inequality would have been reduced throughout the 1982 to 1996 time period.

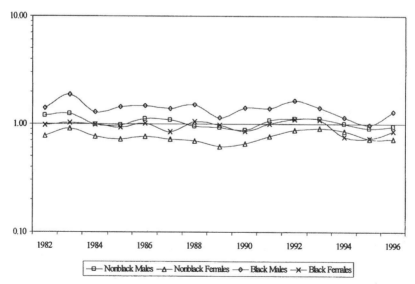

Figure 7.23. Purged odds of unemployment due to quits and job losses (baseline = 0.6).

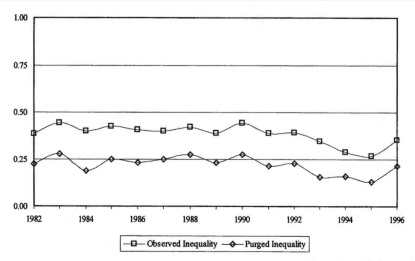

Figure 7.24. Inequality measures for unemployment due to quits and job losses (overall observed inequality = 0.39; overall purged inequality = 0.22).

Unemployed Due to New Entrants and Reentrants. For those unemployed due to being a new entrant or reentrant to the labor market, a comparison of the observed and purged odds in Figs. 7.8 and 7.25 show that, while the baseline odds change very little (0.5 for the observed and 0.4 for the purged),

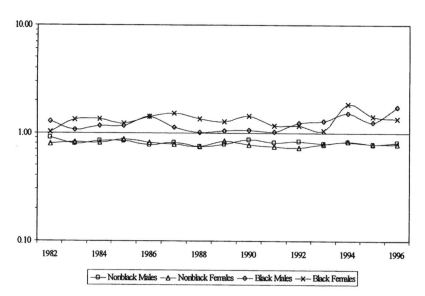

Figure 7.25. Purged odds of new entrant and reentrant unemployment (baseline = 0.4).

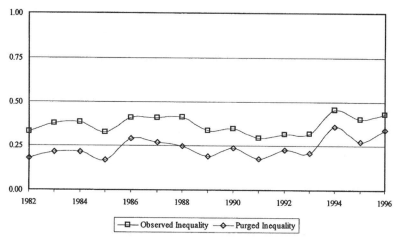

Figure 7.26. Inequality measures for new entrant and reentrant unemployment (overall observed inequality = 0.37; overall purged inequality = 0.25).

once again inequality declines in the purged case relative to the observed case. Here both observed and purged odds tend to fall along racial lines, with nonblacks having lower odds of being an unemployed new entrant or reentrant for both cases. However, the racial gap is clearly closed in the purged odds relative to the observed odds. The inequality measures in Fig. 7.26 suggest that inequality would decline by 32% (= [1 − (0.25/0.37)] × 100) in the purged inequality relative to the observed. The trend in Fig. 7.26 also shows that, had market experiences been independent of race–gender, the racial gap in unemployment due to being a new entrant or reentrant would have shrunk throughout this period.

Not in the Labor Force.   Comparison of the odds in Figs. 7.9 and 7.27 shows that the baseline odds of not being in the labor force declines from an observed odds of 8.1 to a purged odds of 6.4. Moreover, the odds that nonblack women will be not in the labor force noticeably declines in the purged case relative to the observed case. This indicates that, had market experiences been independent of race–gender, nonblack women would have been considerably more likely to be in the labor force relative to the observed case. Though not as dramatic, a similar difference is obtained for black women as well.

Another interesting result here is that, had experiences been independent of race–gender, rather than observing a gender gap in the odds of not being in the labor force, we would have observed a slight racial gap for much of this time frame. That is, the existing relationship between market experiences and race–sex has a tendency to create a gender gap in the likelihood of

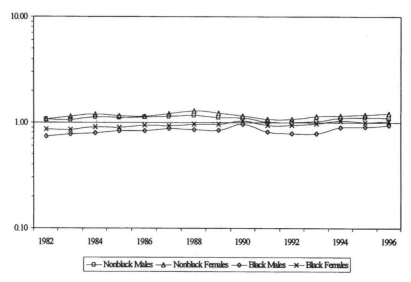

Figure 7.27. Purged odds of not being in the labor foce (baseline = 6.4).

being in the labor force. Were that relationship one of independence, the tendency would be to create a small racial gap in the likelihood of being in the labor force. Thus, these results suggest that eliminating the gender gap by orthogonalizing market experiences across race–gender groups might have the unintended consequence of creating a racial gap.

However, as shown by the inequality measures given in Fig. 7.28, the

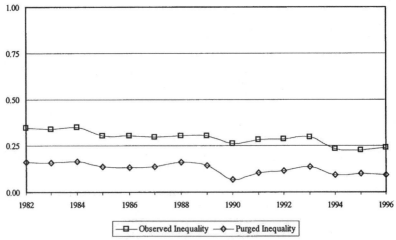

Figure 7.28. Inequality measures for not being in the labor force (overall observed inequality = 0.29; overall purged inequality = 0.13).

racial gap created in this counterfactual world would be less than is the observed gender gap. More precisely, the baseline inequality for not being in the labor force in the counterfactual case would drop by 55% (= [1–(0.13/0.29)] × 100). As the trend in Fig. 7.28 shows, the reduction in inequality holds throughout the 1982 to 1996 time period.

DISCUSSION AND CONCLUSIONS

The major finding that this chapter brings to light is that less than 1% (0.77%) of the total association between labor market experiences and labor force outcomes between 1982 and 1996 can be accounted for by race–gender differences in that relationship. Moreover, temporal shifts in that relationship accounted for less than 1% (0.80%) of the total experience–outcome association as well. This indicates considerable stability in the experience–outcome association over time and across race–gender groups.

Although the experience–outcome relationship is stable, we have shown that race-gender differences in labor force outcomes themselves are due in part to race–gender differences in market experiences. This indicates that the labor market is very much segmented along race and gender lines, and that these lines are very stable and distinct over our time period. Throughout this time period, nonblack men dominate year-round adequate full-time employment. Nonblack women dominate year-round voluntary part-time employment. Black men dominate a pocket of unemployment with prolonged, and unsuccessful, job searches. Black women dominate another pocket of unemployment with shorter, though still unsuccessful, job searches.

These results have important implications for the arguments discussed at the beginning of the chapter indicating preferences on the part of employers to shy away from blacks and women because of the perceived and assumed work histories of these groups. Although our results indicate that black men are indeed far more likely to experience extended periods looking for work, we also find that the *relationship* between market experience in general and subsequent labor force outcomes does not differ across race–gender groups. Those who experience extended search periods have the same (low) odds of finding adequate full-time employment regardless of race or gender. Our results suggest that racial inequality exists during the search period; black men are far more likely to experience extended searches than any other group and it is that experience that weakens their ability to find adequate full-time employment. Although this is consistent with the employer interviews discussed by Wilson (1996), it is not consistent with the idea that it is blacks alone with these extended search histories who receive such treatment by employers.

How, then, can we reconcile our findings of no evidence for differential treatment with previous literature suggesting that such differential treatment clearly exists? To answer this we return to Ridgeway's (1997) theory of categorization. However, here we extend this argument further, suggesting that these differences may be explained by specifying more precisely *when* employers look beyond the categorization process and use work history information. First, extending the theory to race, we would expect that employers are likely to draw negative inferences about productivity and reliability on the basis of an individual being black. As such stereotypes become embedded in employers' culture, blacks are more likely to experience extended periods of job search while unemployed than are nonblacks. This is precisely what we find, and perhaps explains why blacks are more likely to experience extended and unsuccessful job searches.

However, the converse does not necessarily hold true in the case of nonblacks. That is, employers are not likely to draw *positive* inferences about productivity and reliability solely on the basis of an individual being nonblack or, more precisely, a white male. It is at this point that work histories would become important bits of information in sifting through potential hires. Once a worker establishes an intermittent, or worse, work history, it is used by employers to discard undesirable job candidates. Thus, once a worker develops such a history, she or he ends up with the same difficult time finding adequate employment. However, because nonblacks are not subject to the blanket assumptions to which blacks are subject in the categorization process, nonblacks are less likely than blacks to develop these negatively valued work histories. This is precisely what we find, and helps to explain in part why we see race–gender variation in market experiences, but no race–gender variation in the association between experiences and outcomes.

Why then do women, especially nonblack women, end up in part-time work more than any other group (over twice as likely on average)? Perhaps the answer is to be found in the proportion of women already in part-time employment. Cohen, Broshak, and Haveman (1998) find in their study of sex composition and hiring/promotion in the California savings and loan industry that women are more likely to land managerial positions if a relatively high proportion of women are already employed in those jobs. Yet others (Bielby & Bielby, 1984; Rosenfeld & Kalleberg, 1991; Glass & Riley, 1998) show how family constraints and (the absence of) family-friendly policies influence women's employment alternatives and options. Given the constraints of family responsibilities, the categorization process described by Ridgeway (1997), the lack of strong family-friendly policies in the United States, and the number of women already in part-time jobs, women (re)entering the labor market likely view part-time employment as the most feasible choice under these constraints, regardless of their individual preferences for employment in general.

We have also shown that, had nonblack women and black men and women the same labor market experiences as nonblack men in 1996, race–gender inequality in labor force outcomes would have been reduced by about 30%, with the odds of adequate full-time employment increasing from 9.7 to 13.7. When considering population figures in 1996, this increase corresponds to over 16 million more nonblack women, over 800,000 more black men, and almost 2 million more black women who would have been adequately full-time employed. This realignment in market experiences practically means that policy efforts would need to somehow shift nonblack women away from non-employment and part-time employment toward year-round full-time employment. Black women would need to be shifted away from periods of unsuccessful job search and toward year-round full-time employment. Similarly, black men would need to be shifted away from extended periods of unsuccessful job searching and toward year-round full-time employment.

Our results also suggest that there would be unintended negative consequences associated with equalizing labor-market experiences across race–gender groups. Most notably, the odds of low-income full-time employment would likely increase. Also, our results indicate that such efforts would likely do little to change the baseline odds of part-time employment or unemployment. However, for voluntary part-time employment and unemployment due to quits, losses, new entrants to the market, and reentrants to the market, our results suggest that race–gender inequality would be reduced.

On the other hand, given the temporally stable and group-homogeneous nature of the relationship between market experiences and labor force outcomes, our results suggest that attempts to create race–sex equality in labor force outcomes that focus on equalizing that relationship will likely have little noticeable effect at the national level. Consequently, policies based on ensuring the equal treatment of market experiences by employers in the hiring process are not likely to show much impact on race–gender inequities in labor force outcomes. Rather, policies oriented toward race–gender equality in labor force outcomes would likely have a more noticeable impact were they to focus on equality of market experiences, with the goal of moving aggregate market experience distributions toward those of nonblack males. Our results suggest that successful policies along this line would pay off in terms of increased work-force productivity and decreased labor force inequality.

NOTES

1. For diverse approaches see, for example, Becker (1975), Kalleberg and Berg (1987),Wilson (1987, 1996), Bills (1988), Clogg and Eliason (1990), Clogg, Eliason, and Wahl (1990), Eliason (1995), Wilson, Tienda, and Wu (1995), Holzer (1996), and Tam (1997).

2. We use the term *voluntary* here only to refer to economic context. Someone who is classified

as "voluntarily part-time employed" in our analysis has listed no economic reason (e.g., lack of available full-time work or slack work due to economic downturns) for involuntary part-time employment. Readers should keep this in mind because there are other reasons (such as family responsibilities) that would be classified as voluntary part-time employment that may be involuntary.

3. For related discussions see England and Farkas (1986), Wilson (1987, 1996), Kalleberg and Berg (1987), Bills (1988), Reskin and Roos (1990), England (1992), Tomaskovic-Devey (1993), and Holzer (1996).

4. These decompositions are a modest extension of that developed by Becker and Clogg (1989). The Becker–Clogg decomposition is based on one group, whereas this decomposition is based on two groups with one of the groups being a temporal component.

5. This value is obtained by taking the difference in the likelihood ratio statistics for the model and the baseline conditional independence model, and dividing by the baseline model.

6. These component values of 0.77 and 0.80% will generally not add to the combined 1.09% because race–sex group distributions are not independent of time period.

7. Distributions for all years are available from the authors.

8. It is important to note that the results presented here do depend on which group at which time point is considered the standard. We chose the 1996 nonblack male experience distribution because it was the most recent in our series, as well as being one of the more favorable distributions. Also, with the purging method, no standard is necessary in calculating purged frequencies, or functions of them. However, when no standard is used, the tendency is for outcome rates to be shifted toward the middle with respect to the across-group distribution of any one outcome category. In this case we chose not to go in this direction because it lends itself to somewhat perverse results. For example, although more equality is obtained across groups for adequate full-time employment, the baseline odds of adequate full-time employment is less than that observed when no standard is used.

APPENDIX 7.1. MODELS AND METHODS

To model the association between market experiences and labor force outcomes, we use an extension of the RC association model as described by Clogg, Eliason, and Wahl (1990) in their analysis of 1982 market experiences and labor force outcomes. Specifically, we extend the RC-G association model (Clogg, 1982f; Becker & Clogg, 1989) to accommodate differences across race–sex groups and differences over time in the experience–outcome relation. This partitioning into race–sex group and temporal differences is an adaptation of the partitioning developed by Becker and Clogg (1989) for RC-G association models. See Clogg (1982b), Goodman (1986, 1991), Becker (1989, 1990), Becker and Clogg (1989), and Eliason (1995b) for useful developments and discussion of the RC association model.

The RC-G association model for a cross-classification of market experiences (as measured by the variable EXPER with $i = 1, \ldots, 16$ categories) by labor force outcomes (as measured by the variable POSITION with $j = 1, \ldots, 9$ categories) for $k = 1, \ldots, 4$ race–sex groups (nonblack males, black males, nonblack females, and black females) and $t = 1, \ldots, 15$ years (1982 through 1996) can be written

$$\log\left(F_{ij(kt)}\right) = \lambda_{(kt)} + \lambda_{i(kt)}^{E} + \lambda_{j(kt)}^{P} + \sum_{m=1}^{M} \phi_{m(kt)} \mu_{im(kt)} \nu_{jm(kt)} \tag{7.1}$$

$F_{ij(kt)}$ gives the expected frequency under the model for the i-th market experience (EXPER) and the j-th labor-force outcome (POSITION) for race–sex group k at time t, $\lambda_{(kt)}$ fits the k-th race-sex group size at time t, $\lambda_{i(kt)}$ and $\lambda_{j(kt)}$ respectively fit the market experience and labor-force-outcome marginal distributions for the k-th race–sex group at time t, $\phi_{m(kt)}$ is the intrinsic association in latent dimension m between market experiences and labor force outcomes for race–sex group k at time t, $\mu_{im(kt)}$ is the score in latent dimension m for the i-th market experience for race–sex group k at time t, and $\nu_{jm(kt)}$ is the score in latent dimension m for the j-th labor force outcome for race–sex group k at time t.

To identify the latent scores, we constrain the μ's and ν's to have zero mean and unit variance, and to be orthogonal across latent dimensions. For the RC-G model given in Eq. (7.1), these restrictions are given by

$$\sum_{i=1}^{I} g_{i(kt)} \mu_{im(kt)} = \sum_{j=1}^{J} h_{j(kt)} \nu_{jm(kt)} = 0 \tag{7.2}$$

$$\sum_{i=1}^{I} g_{i(kt)} \mu_{im(kt)} \mu_{im'(kt)} = \sum_{j=1}^{J} h_{j(kt)} \nu_{jm(kt)} \nu_{jm'(kt)} = \delta_{mm'} \tag{7.3}$$

where $\sigma_{mm'}$ is 1 when m = m¢ and 0 when m / m'. The $g_{i(kt)}$ and $h_{j(kt)}$ are EXPER and POSITION weights, respectively, for group k at time t. Typical choices for the weights include $g_{i(kt)} = h_{j(kt)} = 1$ (also known as uniform weighting) and $g_{i(kt)} = p_{i+(kt)}$ and $h_{j(kt)} = p_{+j(kt)}$ where the p's are the marginal proportions (also known as marginal weighting). In this setting, we use uniform weights because they result in row and column scores that are independent of row and column marginal distributions, making across-group and over-time comparisons independent of across-group or over-time differences in marginal distributions. This is an attractive property shared with interaction terms in log-linear models. See Becker and Clogg (1989) for relevant discussion.

When $\phi_{m(kt)} = 0$ for all latent dimensions and all groups at all time periods, the RC-G association model in Eq. 7.1 is equivalent to the log-linear model of conditional independence. In this context the conditional independence model is often considered the baseline from which to compare other models. When max $(I - 1, J - 1)$ latent dimensions are estimated for each group at each time period (where I and J are the number of rows and columns, respectively), the RC-G model in Eq. (2.1) is equivalent to the saturated log-linear model.

One way to interpret the RC-G association model is to view the log-odds ratio as a function of the row and column scores. For a model with M latent dimensions, the log-odds ratios for group k at time t are given by

$$\Phi_{ii'jj'(kt)} = \log\left(\frac{F_{ij(kt)}/F_{ij'(kt)}}{F_{i'j(kt)}/F_{i'j'(kt)}}\right) = \sum_{m=1}^{M}\phi_{m(kt)}\left(\mu_{im(kt)}-\mu_{i'm(kt)}\right)\left(\nu_{jm(kt)}-\nu_{j'm(kt)}\right) \quad (7.4)$$

Equation (7.4) shows that the log-odds ratio involving market experiences i and i' and labor force outcomes j and j' is a function of the distance between experience scores i and i' and outcome scores j and j' in dimension m. In this context, the estimated labor force outcome scores may be interpreted as giving the order among, and relative distances between pairs of, labor-force outcomes in dimension m, with respect to the estimated market experience scores in dimension m. The farther apart are any two labor force outcome scores $\nu_{jm(kt)}$ and $\nu_{j'm(kt)}$, the more dissimilar are outcomes j and j' with respect to market experiences. Similarly, the estimated market experience scores in each latent dimension m may be interpreted as giving the order among, and relative distances between pairs of, labor market experiences in dimension m, with respect to the estimated labor force outcome scores in dimension m. The farther apart are any two labor market experience scores $\mu_{im(kt)}$ and $\mu_{i'm(kt)}$ the more dissimilar are experiences i and i' with respect to labor force outcomes.

The $\phi_{m(kt)}$ give the log-odds ratio for any two market experiences and any two labor-force outcomes that are one unit apart on row and column scores. It is in this context that the $\phi_{m(kt)}$ are referred to as the intrinsic association between rows and columns making up the cross-classification (Goodman, 1986, 1991). The larger the intrinsic association in dimension m, the stronger is the association between market experiences and labor-force outcomes as indicated by the set of scores in dimension m. In the saturated RC-G association model for the EXPER by POSITION cross-classification, the intrinsic associations and the latent scores give all of the information about the association between market experiences and labor force outcomes. Placing across-group and/or over-time homogeneity constraints on the estimated intrinsic associations and latent scores allows us to partition that total experience–outcome association into components due to group differences and differences over time.

We adopt the partitioning method developed in Becker and Clogg (1989) for the RC-G association model. Here, however, we provide a modest extension to the Becker–Clogg partitioning to accommodate the across-group and across-time components of our EXPER by POSITION cross-classification. To partition the overall EXPER by POSITION association into these components we consider three constrained models in addition to the unconstrained RC-G association model given in Eq. 7.1. These are the *group-heterogeneity temporal-homogeneity model*

$$\log\left(F_{ij(kt)}\right) = \lambda_{(kt)} + \lambda_{i(kt)}^{E} + \lambda_{j(kt)}^{P} + \sum_{m=1}^{M}\phi_{m(k)}\mu_{im(k)}\nu_{jm(k)} \tag{7.5}$$

the *group-homogeneity temporal-heterogeneity model*

$$\log\left(F_{ij(kt)}\right) = \lambda_{(kt)} + \lambda_{i(kt)}^{E} + \lambda_{j(kt)}^{P} + \sum_{m=1}^{M}\phi_{m(t)}\mu_{im(t)}\nu_{jm(t)} \tag{7.6}$$

and the *group-homogeneity temporal-homogeneity model*

$$\log\left(F_{ij(kt)}\right) = \lambda_{(kt)} + \lambda_{i(kt)}^{E} + \lambda_{j(kt)}^{P} + \sum_{m=1}^{M}\phi_{m}\mu_{im}\nu_{jm} \tag{7.7}$$

For each of these models we estimate, in turn, a one-dimension solution [the RC(1)-G model] through an eight-dimension solution [the RC(8)-G model], the maximum number of latent dimensions allowed for this table. Combined with the conditional independence and unconstrained RC-G models, we estimate a total of 33 models.

The group-heterogeneity temporal-homogeneity model in Eq. (7.5) constrains the intrinsic associations and the market experience and labor force outcomes scores to be equal across time periods, but allows them to vary across race–sex groups. This model thus forces the experience–outcome association to be unchanged over time, but allows them to be different across race–sex groups. This model is nested in the unconstrained model, allowing a decomposition of the likelihood ratio statistic into the proportion of the overall association due to race–sex differences in the EXPER–POSITION association in each of the eight latent dimensions.

The group-homogeneity temporal-heterogeneity model given in Eq. (7.6) constrains the intrinsic associations and the market experience and labor force position scores to be equal across race–sex groups, but allows them to vary over time. This model thus forces the experience–outcome association to be the same across race–sex groups, but allows them to change over time. This model is also nested in the unconstrained model, allowing a decomposition of the likelihood ratio statistic into the proportion of the overall association due to differences in the EXPER–POSITION association over time in each of the eight latent dimensions.

Appendix Table 7.2. Partitioning of the Likelihood Ratio Statistic into Race–Sex Group and Time Period Homogeneity and Heterogeneity.

Model	Race–sex homogeneity Temporal homogeneity	Race–sex homogeneity Temporal heterogeneity	Race–sex heterogeneity Temporal homogeneity	Race–sex heterogeneity Temporal heterogeneity
Likelihood ratio statistic				
Cond. Indep.	1,990,852 (7020)	—	—	—
RC(1)	300,265 (6998)	297,865 (6990)	293,805 (6932)	289,149 (5700)
RC(2)	116,542 (6978)	111,058 (6390)	104,664 (6852)	95,272 (4500)
RC(3)	29,405 (6960)	24,780 (6120)	23,659 (6780)	13,372 (3420)
RC(4)	24,469 (6944)	19,948 (5880)	19,791 (6716)	7,347 (2460)
RC(5)	23,881 (6930)	17,833 (5670)	17,298 (6660)	2,826 (1620)
RC(8)	21,658 (6900)	15,249 (5220)	15,855 (6540)	0 (0)
Likelihood ratio statistic decomposition				
Dimension 1	1,690,587 (22)	1,692,987 (330)	1,697,047 (88)	1,701,703 (1320)
Dimension 2	183,723 (20)	186,807 (300)	189,141 (80)	193,877 (1200)
Dimension 3	87,137 (18)	86,278 (270)	81,005 (72)	81,900 (1080)
Dimension 4	4,936 (16)	4,832 (240)	3,868 (64)	6,025 (960)
Dimension 5	588 (14)	2,115 (210)	2,493 (56)	4,521 (840)
Dimension 6--8	2,223 (30)	2,584 (450)	1,443 (120)	2,826 (1620)
Heterogeneity	21,658 (6900)	15,249 (5220)	15,855 (6540)	—

Percent decomposition

Dimension 1	84.92%	85.04%	85.24%	85.48%
Dimension 2	9.23%	9.38%	9.50%	9.74%
Dimension 3	4.38%	4.33%	4.07%	4.11%
Dimension 4	0.25%	0.24%	0.19%	0.30%
Dimension 5	0.03%	0.11%	0.13%	0.23%
Dimension 6–8	0.11%	0.13%	0.07%	0.14%
Heterogeneity	1.09%	0.77%	0.80%	—

Note: The first entry is the likelihood ratio statistic and the second entry in parentheses is the degrees of freedom. Note also that the degrees of freedom are affected by the presence of three structural zeros in each experience-by-position cross-classification.

APPENDIX 7.3. A METHOD FOR CALCULATING INFLUENCE MEASURES

One of the most attractive features of the RC and RC-G association models is that functions of the intrinsic associations and latent scores lend themselves nicely to geometric interpretations of the underlying complex relation in the cross-classification under study (Goodman, 1986, 1991; Clogg, Eliason, & Wahl 1990; Clogg & Eliason, 1990; Eliason, 1995b). At the heart of these interpretations are the distances and correlations between rows and columns. We also introduce here another measure that combines distances and correlations into one summary value, one that gives the relative influence that any one row has on any one column. This influence measure, though derived from the underlying geometry of the RC association model, can be shown to be the sum of the interaction terms in the model. Thus, this influence measure is readily interpretable from the standpoint of the modeled log-odds ratios, as well as from the underlying geometry of the model.

Distances between Points

Perhaps the most common measure derived from the geometry of the RC association model to determine the relation between a row and a column is the Euclidean distance between a row point and a column point. Note that distances between rows and/or columns generally cannot be represented by distances between corresponding row and/or column scores alone. As shown in previous work (Goodman, 1991, Eliason, 1995b), distances between rows and/or columns are best obtained using adjusted scores. These adjusted scores, μ^*_{mi} and v^*_{mj}, are given by

$$\mu^*_{mi(kt)} = \phi_m^{\alpha_m} \mu_{mi(kt)} \tag{7.8}$$

$$v^*_{mj(kt)} = \phi_m^{(1-\alpha_m)} v_{mj(kt)} \tag{7.9}$$

where $0 < \alpha_m < 1$. Without otherwise compelling reasons, most often $\alpha_m = 0.5$ for all M dimensions. This is the value we use in the text.

Given the adjusted scores in Eqs. (7.8) and (7.9), and assuming the orthogonality constraints hold in Eq. (7.3), the distance between market experience i and labor force outcome j for race–sex group k at time t may be given by

$$d_{ij(kt)} = \sqrt{\sum_m \left(\mu^*_{im(kt)} - v^*_{jm(kt)}\right)^2} \tag{7.10}$$

This measure is readily interpretable; points closer together, indicated by a relatively small $d_{ij(kt)}$, are more strongly related to one another than points farther apart, as indicated by a relatively large $d_{ij(kt)}$.

This measure, however, is not without its drawbacks. Consider two points, one representing some market experience and another representing some labor force outcome, that are close to the origin. These two points, which will register a relatively small $d_{ij(kt)}$ simply by virtue of their being close to the origin, may in fact be completely unrelated. Thus, some prefer instead to assess the relationship between pairs of points by the correlation between them.

Correlations between Points

To remedy the problem associated with the distance measure just mentioned, another summary measure is sometimes preferred to describe the relation between any row–column pair. This is the correlation between any two points in the system. Let $U^*_{i(kt)} = [\mu^*_{i1(kt)}, \ldots, \mu^*_{iM(kt)}]'$ be a vector of adjusted row scores uniquely defining market experience i for race–sex group k at time t. Similarly, let $V^*_{j(kt)} = [v^*_{j1(kt)}, \ldots, v^*_{jM(kt)}]'$ be a vector of adjusted column scores uniquely defining labor-force outcome j for race–sex group k at time t. The correlation between market experience i and labor force position j for group k at time t is then given by

$$r_{ij(kt)} = \cos\left(\angle U^*_{i(kt)} V^*_{j(kt)}\right) = \frac{U^*_{i(kt)} \cdot V^*_{j(kt)}}{\left\|U^*_{i(kt)}\right\|\left\|V^*_{j(kt)}\right\|} \tag{7.11}$$

where the intermediate result gives the cosine of the angle formed by the two vectors $U^*_{i(kt)}$ and $V^*_{j(kt)}$ subtended at the origin. The numerator in the final result is the dot product between, and the denominator gives the product of the magnitudes of, vectors $U^*_{i(kt)}$ and $V^*_{j(kt)}$.

With reference to the geometry of the RC-G association model, a negative (positive) correlation indicates that positive movement along vector $U^*_{i(kt)}$ relates to a corresponding negative (positive) movement along vector $V^*_{j(kt)}$. With reference to our substantive context, the correlation $r_{ij(kt)}$ between market experience i and labor-force outcome j gives the direction and magnitude of the relationship between i and j. A negative correlation indicates that as one is more likely to have had market experience i, the less likely she or he is to attain labor-force outcome j. A positive correlation indicates that as one is more likely to have had market experience i, the more likely she or he is to attain labor force outcome j. A correlation of zero indicates that there is no relationship between market experience i and labor-force outcome j.

The correlation, as with the distance, is not without its shortcomings. The drawback to the correlation is that it does not tell us anything about the relative proximity of any two points. That is, two points could be considerably far apart and yet the correlation, if the angle between the corresponding vectors defined above were as such, could easily indicate quite a strong association. It is for this reason that we define a third measure of the relationship between market experience i and labor-force outcome j, one that combines the information given in the distance and correlation measures.

Combining Distances and Correlations

Both distances and correlations between a row point and a column point in the M-dimensional space can be useful measures in describing the row–column relationship in the two-way cross-classification. However, both of these measures, as noted above, contain only partial information about that relationship. It would be useful, then, to consider a measure that combines information from both into one parsimonious summary of the association between i and j. Toward that end, we borrow a measure intended to gauge the amount of work required for one object to move another, an idea common to the physical sciences, but one that lends itself nicely to our purposes as well.

First, consider the question as posed by physical scientists, phrased in terms of the geometry of the problem, regarding how much effort must be applied along vector $U^*_{i(kt)}$ to effect movement along vector $V^*_{(jt)}$. Considering that the vectors $U^*_{i(kt)}$ and $V^*_{j(kt)}$ uniquely determine market experience i and labor force outcome j respectively, this question may be rephrased in the language of our problem: How much influence does market experience i have on attaining labor force position j? This measure of work between two objects as commonly used in the physical sciences, or the measure of influence of market experience i on labor force position j as recast for our context, is defined as

$$w_{ij(kt)} = \left\|U^*_{i(kt)}\right\| \cos\left(\angle U^*_{i(kt)} V^*_{j(kt)}\right) \left\|V^*_{j(kt)}\right\| \tag{7.12}$$

The result in Eq. (7.12), when combined with the result in Eq. (7.11), reduces to

$$w_{ij(kt)} = U^*_{i(kt)} \cdot V^*_{j(kt)} \tag{7.13}$$

which is the dot product between vectors $U^*_{i(kt)}$ and $V^*_{j(kt)}$.

By replacing the vectors in Eq. (7.13) with their corresponding elements, we obtain the following result relating $w_{ij(kt)}$ to the interaction terms in the RC-G association model,

$$w_{ij(kt)} = \sum_{m=1}^{M} \mu^{*}_{im(kt)} v^{*}_{jm(kt)} = \sum_{m=1}^{M} \phi_{m(kt)} \mu_{im(kt)} v_{jm(kt)} \tag{7.14}$$

Equation (7.14) shows that the measure of influence between market experience i and labor force outcome j for group k at time t is equal to the row–column interaction component of the RC-G association model. Although this is the form for the unconstrained RC-G model, influence measures for the constrained RC-G models discussed in the text are readily obtained by substituting into Eq. (7.14) the interactions from the constrained models.

When $w_{ij(kt)} = 0$, market experience i and labor force outcome j are unrelated in race–sex group k at time t. When $w_{ij(kt)} < 0$, market experience i and labor force position j are negatively related. When $w_{ij(kt)} > 0$, market experience i and labor force position j are positively related. As with the correlation discussed above, a negative (positive) relationship indicates that as one is more likely to have had market experience i the less (more) likely one is to attain labor-force outcome j.

Along with the sign of the $w_{ij(kt)}$, the magnitude is also readily interpretable. First note that the $w_{ij(kt)}$ sum to zero row-wise, column-wise, and overall. That is,

$$\sum_{i} w_{ij(kt)} = \sum_{j} w_{ij(kt)} = \sum_{i,j} w_{ij(kt)} = 0 \tag{7.15}$$

From this property, the corresponding means of the $w_{ij(kt)}$ are also zero. Second, the overall sum of squares is given by

$$\sum_{i,j} w^{2}_{ij(kt)} = \sum_{m} \phi^{2}_{m(kt)} \tag{7.16}$$

Thus, the mean and variance for the $I \times J$ set of $w_{ij(kt)}$ for race–sex group k at time t are given by

$$\bar{w}_{(kt)} = 0$$
$$s^{2}_{w_{(kt)}} = \frac{\sum_{m} \phi^{2}_{m(kt)}}{I \times J} \tag{7.17}$$

These results may be used to describe the relative importance of the magnitude of the mutual influence for any market experience i and labor-force outcome j. A generally useful rule, one that we apply here, is that an (i, j) pair for which the estimated $\left| \hat{w}_{ij(kt)} \right| > 2 s_{w_{(kt)}}$ can be considered to have relatively high mutual influence relative to the $I \times J$ set of $w_{ij(kt)}$ for race–sex group k at time t.

Note that this guideline is purely descriptive with respect to the sampled cross-classification. That is, it allows one to compare the relative influence for any (i, j) pair against the overall set of row–column influences. From the standpoint of the overall set of row–column influences, those (i, j) pairs for which $\left| \hat{w}_{ij(kt)} \right| > 2 s_{w_{(kt)}}$ will have a relatively high contribution in moving the data away from row–column independence.

One may wish to consider, on the other hand, the hypothesis of no influence between row i and column j in the population. In that case the ratio $z^* = \hat{w}_{ij(kt)} \Big/ \mathrm{ase}\left(\hat{w}_{ij(kt)} \right)$, where $\mathrm{ase}\left(\hat{w}_{ij(kt)} \right)$ is the asymptotic standard error of $w_{ij(kt)}$, may be used in large samples. Under the null hypothesis of no influence between row i and column j in the population, z^* will have a standard unit normal distribution and standard inference tests may be used to assess that null hypothesis. It is important to keep in mind, however, that this inferential question is distinct from the descriptive question considered above and in the text.

Appendix Table 7.4. Annual Percentage Distributions of Labor Market Experiences by Race-Sex Groups

	Reagan							Bush				Clinton			
	1982	1983	1984	1985	1986	1987	1988	1989	1990	1991	1992	1993	1994	1995	1996
Nonblack males (%)															
Nonworkers															
Not looking	17.90	18.50	19.24	19.18	19.28	19.22	19.28	19.08	18.97	19.27	19.97	20.33	20.65	20.54	20.90
Looked 15+ weeks	0.64	1.27	1.26	0.78	0.56	0.61	0.49	0.40	0.35	0.48	0.76	0.88	0.96	0.66	0.61
Looked 1–14 weeks	0.46	0.51	0.48	0.37	0.31	0.31	0.31	0.24	0.26	0.30	0.36	0.39	0.50	0.47	0.44
Part year–Part time															
Not looking	4.70	4.42	4.32	4.43	4.38	4.42	4.76	4.82	4.55	4.45	4.16	4.17	4.38	4.52	4.41
Looked 15+ weeks in 1 stretch	0.65	0.87	0.78	0.55	0.47	0.49	0.44	0.34	0.33	0.47	0.59	0.67	0.63	0.50	0.58
Looked 15+ weeks in 2+ stretches	0.74	0.93	0.87	0.67	0.57	0.63	0.57	0.45	0.47	0.55	0.75	0.77	0.62	0.52	0.49
Looked 1–14 weeks in 1 stretch	0.79	0.93	0.85	0.76	0.77	0.75	0.64	0.59	0.64	0.68	0.74	0.70	0.55	0.70	0.55
Looked 1–14 weeks in 2+ stretches	0.48	0.45	0.44	0.44	0.47	0.37	0.39	0.43	0.39	0.44	0.39	0.46	0.35	0.37	0.37
Part year–Full time															
Not looking	6.35	5.79	5.89	6.04	6.15	6.11	6.21	6.30	6.48	5.98	5.80	5.39	5.52	5.77	5.50
Looking 15+ weeks in 1 stretch	2.57	3.81	3.22	2.41	2.41	2.48	2.25	1.62	1.58	2.08	2.64	2.77	2.63	2.16	1.95
Looking 15+ weeks in 2+ stretches	2.27	2.55	2.14	1.95	1.75	1.69	1.35	1.33	1.20	1.41	1.68	1.61	1.44	1.16	1.03
Looking 1–14 weeks in 1 stretch	4.06	3.86	3.61	3.52	3.51	3.44	3.14	2.84	3.09	3.32	3.40	3.14	2.55	2.62	2.40
Looking 1–14 weeks in 2+ stretches	1.88	1.82	1.49	1.50	1.53	1.33	1.26	1.38	1.30	1.47	1.28	1.25	1.10	0.98	0.88
Full year															
Voluntary part time	3.06	3.01	2.99	2.85	3.00	2.92	2.98	3.32	3.40	3.22	3.21	3.23	3.40	3.20	3.46
Involuntary part time	0.70	0.67	0.92	0.83	0.82	0.83	0.75	0.63	0.73	0.81	1.00	1.02	1.01	0.93	0.81
Full time	52.76	50.61	51.50	53.71	54.02	54.41	55.20	56.24	56.27	55.08	53.30	53.23	43.70	54.93	55.62
Nonblack females (%)															

(Continued)

Appendix Table 7.4. Continued

	Reagan							Bush				Clinton			
	1982	1983	1984	1985	1986	1987	1988	1989	1990	1991	1992	1993	1994	1995	1996
Nonworkers															
Not looking	40.94	41.21	40.54	39.41	39.20	38.54	37.73	36.89	36.89	36.73	36.88	36.97	36.26	36.08	35.89
Looked 15+ weeks	0.49	0.78	0.77	0.57	0.46	0.45	0.36	0.26	0.18	0.22	0.45	0.57	0.77	0.57	0.46
Looked 1–14 weeks	0.92	1.11	0.99	0.86	0.77	0.65	0.54	0.53	0.46	0.57	0.52	0.56	0.65	0.71	0.63
Part year–Part time															
Not looking	9.36	8.50	8.68	9.30	9.13	9.03	9.11	9.19	9.14	8.63	8.31	7.96	8.41	8.53	8.32
Looked 15+ weeks in 1 stretch	0.74	0.87	0.78	0.63	0.67	0.61	0.51	0.52	0.45	0.49	0.65	0.72	0.75	0.62	0.63
Looked 15+ weeks in 2+ stretches	0.42	0.64	0.48	0.47	0.40	0.47	0.39	0.30	0.30	0.38	0.49	0.42	0.42	0.41	0.39
Looked 1–14 weeks in 1 stretch	1.33	1.37	1.31	1.29	1.29	1.17	1.13	1.06	0.95	1.16	1.17	1.10	0.88	0.89	0.98
Looked 1–14 weeks in 2+ stretches	0.59	0.53	0.45	0.48	0.46	0.43	0.44	0.40	0.42	0.48	0.44	0.42	0.39	0.35	0.38
Part year–Full time															
Not looking	6.98	6.17	6.23	6.89	6.61	6.53	6.78	6.66	7.34	6.66	6.07	5.50	5.93	6.00	6.30
Looking 15+ weeks in 1 stretch	1.35	1.61	1.45	1.16	1.18	1.16	0.90	0.99	0.93	1.08	1.19	1.34	1.31	1.19	1.16
Looking 15+ weeks in 2+ stretches	0.77	0.78	0.66	0.57	0.55	0.56	0.45	0.38	0.45	0.52	0.48	0.44	0.48	0.41	0.35
Looking 1–14 weeks in 1 stretch	2.59	2.38	2.14	2.34	2.36	2.19	1.96	1.84	2.04	2.28	1.98	1.78	1.63	1.78	1.66
Looking 1–14 weeks in 2+ stretches	0.87	0.82	0.65	0.64	0.64	0.66	0.60	0.56	0.61	0.61	0.50	0.46	0.39	0.48	0.42
Full year															
Voluntary part time	6.05	6.31	6.63	6.33	6.46	6.84	7.27	7.50	7.11	7.25	7.15	7.27	7.72	8.02	7.69
Involuntary part time	0.92	1.23	1.12	1.16	1.17	1.17	1.17	1.07	1.01	1.03	1.30	1.53	1.39	1.17	1.08
Full time	25.70	25.69	27.13	27.91	28.66	29.52	30.65	31.87	31.72	31.93	32.44	32.95	32.62	32.78	33.68

Black males (%)															
Nonworkers															
Not looking	26.42	28.28	26.85	27.39	25.81	26.47	27.00	26.66	27.00	26.87	27.46	28.02	28.40	28.64	27.86
Looked 15+ weeks	3.37	4.07	4.41	3.47	2.23	1.85	1.94	1.29	1.54	1.47	2.02	2.24	2.58	1.88	2.90
Looked 1–14 weeks	1.16	1.50	1.67	1.06	1.01	1.15	0.75	0.93	0.77	1.12	0.94	1.02	1.30	1.23	0.99
Part year–Part time															
Not looking	3.90	3.31	3.40	3.63	3.79	3.62	4.39	4.51	4.49	3.52	4.69	4.18	4.63	4.18	3.99
Looked 15+ weeks in 1 stretch	1.29	1.28	1.40	1.31	1.05	1.07	1.13	0.85	0.93	0.67	0.87	1.18	1.00	0.98	0.96
Looked 15+ weeks in 2+ stretches	1.11	1.93	1.57	1.95	1.11	1.69	1.01	1.37	0.90	1.10	1.39	1.23	0.95	0.97	0.55
Looked 1–14 weeks in 1 stretch	0.64	1.07	1.27	0.74	1.25	0.95	1.01	0.88	0.92	0.59	0.79	0.87	0.78	0.55	0.48
Looked 1–14 weeks in 2+ stretches	0.82	0.63	0.93	0.56	0.48	0.58	0.58	0.39	0.76	0.53	0.42	0.42	0.60	0.42	0.44
Part year–Full time															
Not looking	4.34	4.48	4.60	4.92	4.74	5.91	5.14	5.78	5.31	6.05	5.30	5.84	5.36	4.27	4.76
Looking 15+ weeks in 1 stretch	3.90	5.46	4.41	4.30	3.26	3.07	3.05	2.78	2.61	2.93	4.06	4.10	3.66	3.08	2.87
Looking 15+ weeks in 2+ stretches	2.86	3.42	2.61	2.57	2.93	2.07	2.50	2.24	1.96	2.90	3.28	1.82	1.93	1.35	1.66
Looking 1–14 weeks in 1 stretch	4.79	3.74	3.01	3.09	3.48	3.56	3.46	3.12	3.54	2.90	3.28	3.35	2.23	2.42	2.33
Looking 1–14 weeks in 2+ stretches	1.89	1.70	1.34	1.29	1.36	1.75	1.26	1.23	1.29	1.22	1.47	1.37	1.17	0.73	1.18
Full year															
Voluntary part time	1.66	1.43	1.97	1.90	2.26	1.60	1.77	2.24	1.99	2.29	2.36	2.11	1.65	2.83	2.31
Involuntary part time	0.70	1.41	1.34	1.64	1.33	1.15	0.78	1.06	1.05	1.55	1.34	1.24	1.00	1.18	1.13
Full time	41.16	36.30	38.91	40.52	43.90	43.49	44.22	44.67	44.95	44.48	41.66	41.02	42.76	45.28	45.58

(Continued)

Appendix Table 7.4. Continued

			Reagan						Bush				Clinton		
	1982	1983	1984	1985	1986	1987	1988	1989	1990	1991	1992	1993	1994	1995	1996
Black females (%)															
Nonworkers															
Not looking	40.24	39.78	40.56	38.07	38.15	37.28	37.95	36.90	36.46	37.64	37.69	38.64	35.52	35.13	36.07
Looked 15+ weeks	2.22	2.95	3.08	2.29	1.89	2.28	1.83	1.33	1.13	1.07	1.29	1.70	2.20	1.94	1.64
Looked 1–14 weeks	3.30	3.11	2.99	2.52	2.07	1.93	1.98	1.66	1.50	1.43	1.68	1.31	1.91	.130	1.30
Part year–Part time															
Not looking	5.32	5.28	5.14	6.23	5.39	5.34	5.09	5.99	5.80	5.35	5.52	5.01	5.58	6.74	5.11
Looked 15+ weeks in 1 stretch	1.02	1.52	1.32	0.91	1.14	1.17	1.13	0.82	0.72	0.94	1.07	0.75	1.34	0.93	0.87
Looked 15+ weeks in 2+ stretches	0.81	1.00	0.87	0.95	1.05	0.59	0.67	0.69	0.73	0.71	0.80	0.85	0.86	0.58	
Looked 1–14 weeks in 1 stretch	1.32	0.94	0.78	1.18	1.15	1.25	1.04	0.92	0.92	0.97	0.93	0.94	0.81	0.79	0.62
Looked 1–14 weeks in 2+ stretches	0.63	0.59	0.57	0.36	0.51	0.54	0.27	0.69	0.48	0.66	0.43	0.28	0.56	0.47	0.41
Part year–Full time															
Not looking	5.57	5.16	4.76	5.17	5.49	6.04	6.31	6.39	6.83	6.99	6.09	6.14	6.25	5.84	5.53
Looking 15+ weeks in 1 stretch	2.33	2.63	2.12	1.83	2.41	2.39	2.23	1.69	1.52	1.97	1.63	2.25	1.91	1.86	2.16
Looking 15+ weeks in 2+ stretches	1.45	1.52	0.92	1.08	1.18	1.16	1.09	0.83	0.76	0.87	0.79	0.79	0.89	0.88	0.59
Looking 1–14 weeks in 1 stretch	3.11	2.64	1.84	2.76	2.37	2.38	2.34	1.71	2.69	2.04	2.35	2.02	2.17	1.85	2.03
Looking 1–14 weeks in 2+ stretches	1.02	0.96	0.85	0.74	0.86	0.94	0.68	1.04	1.05	1.01	0.60	0.63	0.72	0.52	0.56
Full year															
Voluntary part time	2.95	2.80	3.76	3.44	3.28	3.52	3.73	3.20	4.31	4.02	4.34	3.64	3.58	4.16	4.15
Involuntary part time	1.95	1.71	2.06	1.83	1.89	1.71	1.44	1.41	1.13	1.32	1.52	1.81	1.49	1.15	1.28
Full time	26.74	27.40	28.39	30.63	31.26	31.05	32.30	34.56	34.02	33.00	33.36	33.27	34.22	35.59	37.10

APPENDIX 7.5. CALCULATING INEQUALITY KAPPAS

To obtain a precise measure of the overall race–sex inequality in the purged and observed labor force outcome distributions, we use a measure similar to that discussed in Goodman (1991) and used in Hout et al. (1995). Though used by Hout et al. (1995) in assessing the relation between social class and voting, this measure, called kappa (κ) and calculated here as the standard deviation of group differences in market outcomes, is useful in our case as well. In our case this measure, calculated as the standard deviation of race–sex group differences in market outcomes, gives a summary of inequality in market outcomes across race–sex groups over time.

The inequality κ can be written

$$\kappa_t = \sqrt{\frac{1}{JK} \sum_{j,k} \left(\lambda_{jk}^{PG} + \lambda_{jkt}^{PGT} \right)^2} \tag{7.18}$$

where the λ parameters are first obtained for the saturated log-linear model fit to the observed labor force outcome (P superscript) by race–sex group (G superscript) by time period (T superscript) cross-classification. This gives the overall inequality κ_t for the observed relationship at time t, or k_t. We then obtain the λ parameters for the saturated log-linear model estimated on the purged frequencies for the labor force position by race–sex group by time period cross-classification. This gives an overall inequality k_t for the purged relationship at time t, or κ_t. This allows us to compare how much change in overall across-group inequality in market outcomes would be obtained if market experiences were equalized across race–sex groups in the manner discussed above.

We also calculate an inequality κ_{jt} for each of the nine labor-force outcomes,

$$\kappa_{jt} = \sqrt{\frac{1}{K} \sum_{k} \left(\lambda_{jk}^{PG} + \lambda_{jkt}^{PGT} \right)^2} \tag{7.19}$$

As with the overall inequality κ_t, κ_{jt} is obtained for both the observed and purged relationships. This measure gives us the amount of race–sex inequality for each labor force position at time t. Comparison of the measures obtained for the observed and purged frequencies will allow us to show which labor force positions are more affected by orthogonalizing market experiences across race–sex groups.

8

Occupations, Labor Markets, and the Relationship between Labor-Market Experiences and Labor-Force Outcomes

INTRODUCTION

In this chapter we look at occupational differences in the relationship between prior labor market experiences and current labor force outcomes. This is the second dimension that social stratification theorists and policy analysts have detected or suspected changes in the function of labor markets over the past 20 years.

Research on economic restructuring has suggested that the basic occupational structure of the labor market has shifted from a focus on skilled and semiskilled manufacturing occupations and toward service and professional occupations that require serious investments in formal education (e.g., Bluestone & Harrison 1982; *New York Times,* 1996). Numerous interesting questions regarding the basic structure of labor markets arise from this work. Chief among these is the question of whether labor market transitions are destabilized as the labor market changes, or whether economic restructuring manifests itself in the labor market through compositional shifts in the relative size of groups in different labor market categories.

Two examples may help to highlight the different implications of these two dimensions of change. We introduced Joe in the first chapter of the book.

Joe's job as a steelworker was eliminated and he was laid off from U.S. Steel in 1983. Jobs like Joe's used to be the backbone and epitome of the middle class American dream; a steady job with long-term prospects for security, benefits, and pay. If the Joes of this world have a good deal of trouble finding other full-time, year-round jobs (as Joe did), then that could reflect a basic change in the structure of the labor market for workers like him. If Joe's job is completely eliminated, Joe cannot find another full-time job, and he accepts a part-time job, then this might reflect a compositional shift. Joe's full-time job is eliminated and replaced by a job in another labor market niche.

Using data from the 1982–1996 March Current Population Surveys, we find that the relationship between prior labor-market behavior and current labor force position has not changed substantially across occupations. But, as we described in Chapter 7 for race and gender groups, there are substantial differences across occupations in prior labor market experiences. The problem of occupational stratification or "labor market segmentation" is manifested in the ability of specific occupations to provide long-term, stable work histories.

Descriptive Results for Labor Force Outcomes

To facilitate large enough within-cell sample sizes for our market experience by labor force position cross-classification analysis, we group occupations into seven broad categories: *upper nonmanual, sales, administrative support, services, skilled manual, unskilled manual,* and *farm/forestry/fishing occupations.*[1]

Figure 8.1 displays the odds of adequate full-time employment, with a baseline odds of 14.2.[2] Perhaps not surprisingly, those in upper nonmanual occupations enjoy overwhelmingly high odds of having adequate full-time employment throughout the time period. At times (1988 and 1989), those in upper nonmanual occupations have as much as two times the odds of adequate full time employment as those in the next highest occupation, administrative support. The next cluster of occupations, including administrative support, skilled and unskilled manual, and sales, are trending slightly upwards, but never stray far from the baseline odds. Clearly, service and farm/forest/fishing occupations fare the worst here, with odds consistently below the baseline and with only a very slight upward trend.

Low income full-time employment odds are given in Fig. 8.2. Here service occupations are consistently at or near the top of the group. In scattered years (1983, 1984, 1985, 1995, and 1996), sales occupations are quite close to the top as well. Throughout this period unskilled manual and farm, forestry and fishing occupations have the lowest odds of low income full-time employment. Figure 8.2 highlights an important observation made by other researchers in different contexts. Service occupations consistently produce a disproportionate share of poorly paid full-time work (see Bluestone & Harrison, 1982, 1986).

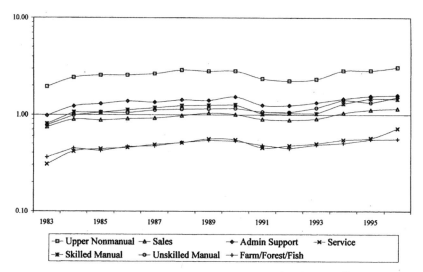

Figure 8.1. Observed odds of adequate full-time employment (baseline = 14.2).

Figure 8.3 shows a separation into two primary groups of occupations in the odds of voluntary part-time employment: Upper nonmanual, sales, administrative support, and service occupations consistently have the highest odds and skilled manual, unskilled manual, and farm, forestry, and fishing occupations consistently have the lowest odds. This division looks the most

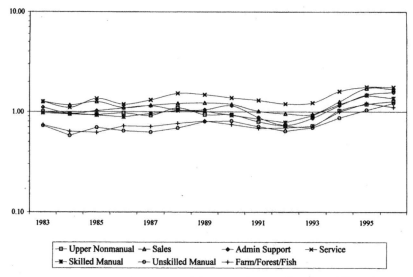

Figure 8.2. Observed odds of low-income full-time employment (baseline = 1.9).

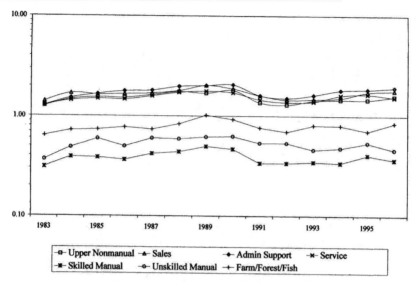

Figure 8.3. Observed odds of voluntary part-time employment (baseline = 2.3).

like a classic manual–nonmanual split. This result also supports those who claim that shifts from manufacturing employment move people away from full-time work.

As shown in Fig. 8.4, part-time employment for economic reasons (such as slack work) does not divide occupations in the same manner as does volun-

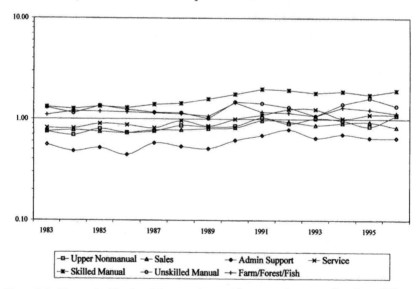

Figure 8.4. Observed odds of part-time employment for economic reasons (baseline = 0.5).

tary part-time employment. From 1987 to 1996, skilled manual occupations have the highest odds, with a generally increasing trend over that time period. For the entire 1983 to 1996 period, administrative support consistently has the lowest odds of part-time employment due to economic reasons.

The odds of locating a part-time job because full-time work was not available is presented in Fig. 8.5. Service occupations are the hardest hit here, except in 1996 where sales and service occupations both have the highest odds. Skilled manual occupations consistently have the lowest odds, suggesting that skilled manual workers have the least difficulty finding work at full-time hours. Through 1992 to 1996, most occupations show a slight decline in these odds.

As with the odds of voluntary part-time employment, Fig. 8.6 suggests that the odds of layoffs fall clearly along the manual/nonmanual distinction. 1983 (the year following the 1982 recession) was the hardest time for layoffs in manual occupations. Those in unskilled manual occupations were also hit hard in 1996. There is little evidence in this figure of "managerial downsizing" among nonmanual occupations in the 1990s (see Gordon, 1996).

Figure 8.7 suggests that the odds of unemployment due to quits and job losses also produce a manual/nonmanual split, though not as pronounced as with layoffs. This distinction starts to break up in 1996 when the odds of quits and job losses for unskilled manual and farm, forestry, and fishing occupations drop to levels similar to upper nonmanual and sales occupations.

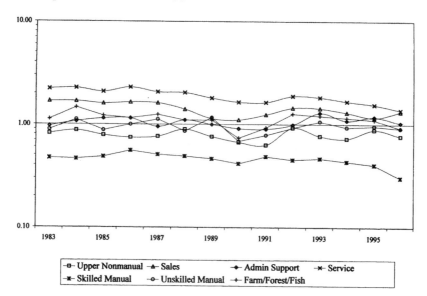

Figure 8.5. Observed odds of part-time employment because no full-time work is available (baseline = 0.8).

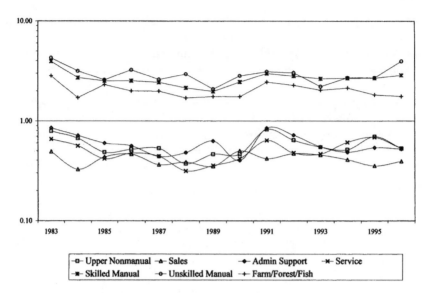

Figure 8.6. Observed odds of unemployment due to layoffs (baseline + 0.2).

Quite different from the other two unemployment states, Fig. 8.8 shows a slight distinction between, on the one hand, upper nonmanual and lower manual, whose odds of new/reentrant unemployment are relatively low compared with, on the other hand, all other occupations. This is especially evident from 1994 onward.

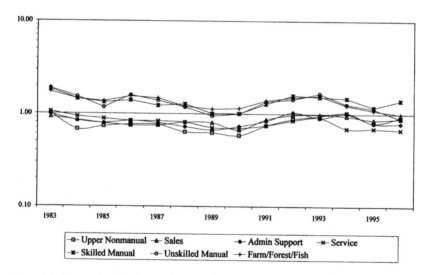

Figure 8.7. Observed odds of unemployment due to quits and job losses (baseline = 0.8).

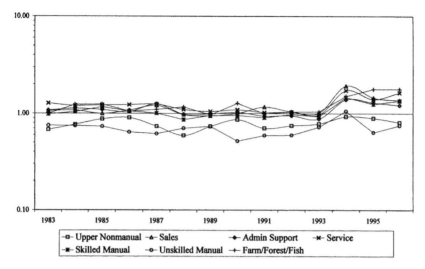

Figure 8.8. Observed odds of new entrants and reentrant unemployment (baseline = 0.4).

Figure 8.9 displays a modest manual/nonmanual split in being out of the labor force, with manual occupations least likely to be associated with being out of the labor force. But by 1994, however, all occupations tend to coalesce at all-time low odds of not being in the labor force.

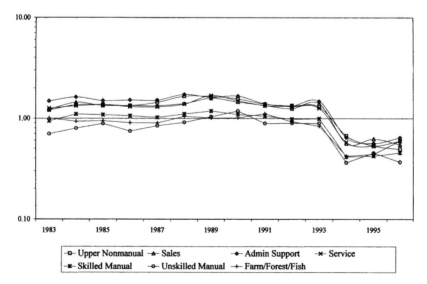

Figure 8.9. Observed odds of not being in the labor force (baseline = 0.8).

ANALYZING THE EXPERIENCE–OUTCOME RELATIONSHIP ACROSS OCCUPATIONS

In this section we present results for the market experience and labor-force outcome associations, looking at the effects of shifts over time and shifts across occupations in this relationship. Then (as we did in Chapter 7 for race and gender groups) we estimate what labor-force outcomes would look like if prior labor-market experiences were equalized across occupations. Estimating this counterfactual scenario allows us to artificially manipulate one important source of stratification in market outcomes and to thereby assess how much of a change in labor force outcomes may be produced by equalizing prior labor market experiences across occupations. In this section we use the same techniques and models that we did in Chapter 7, except here the occupational groups play the role that race–sex groups played in the prior chapter.

Table 8.1 gives the likelihood ratio statistics, decompositions, and percent decompositions across occupations and time for the RC association models described in the previous chapter. Similar to the results for race–sex groups, the association between market experience and labor-force outcomes varies very little across occupations and over time. Only about 3% of the total association is attributable, separately, to changes across occupations and time. Further, anything beyond the third latent dimension accounts for less than 1% of the total association. Therefore, the 1983 to 1996 across-occupation association between market experiences and labor-force outcomes can be described quite well (i.e., to within about 5% of the total association) by the three dimensional RC(3) occupational-homogeneity and temporal-homogeneity model.

Table 8.2 gives the influence measures (described above) for the associations obtained from the three-dimensional model. As in Chapter 7, influence measures greater than two times the standard deviation are shown in bold. These are the cells that have the strongest influence in the table.

A number of the results in Table 8.2 mirror those found for race and sex groups in Chapter 7. Nonworkers who are not looking for work have a strong tendency to not be in the labor force one year later. There is a strong tendency for those who are voluntarily part-time employed throughout the year prior to the survey to be voluntarily part-time employed at the time of the survey. Similarly, there is a strong tendency for those full time employed throughout the year prior to the survey to be adequately full time employed at the time of the survey. Those who are involuntarily part-time employed throughout the prior year are not likely to leave the labor force one year later, and those who have been looking for work for 15 or more weeks are unlikely to accept part-time employment voluntarily.

When we look at across-occupational differences, there are additional

Table 8.1

Partitioning of the Likelihood Ratio Statistic into Occupations and Time Period Homogeneity and Heterogeneity

Model	Occupational homogeneity Temporal homogeneity	Occupational homogeneity Temporal heterogeneity	Occupational heterogeneity Temporal homogeneity	Occupational heterogeneity Temporal heterogeneity
Cond. Independence	669,499			
RC(1)	172,444	169,867	164,314	158,592
RC(2)	76,352	71,103	66,694	53,814
RC(3)	35,796	31,665	26,318	12,398
RC(4)	33,663	28,531	24,015	6,375
RC(5)	32,664	27,573	22,957	3,080
RC(8)	30,579	23,727	21,588	0
Likelihood ratio statistic decomposition				
Dimension 1	497,055	499,632	505,185	510,907
Dimension 2	96,092	98,764	97,620	104,778
Dimension 3	40,555	39,438	40,376	41,417
Dimension 4	2,134	3,134	2,303	6,022
Dimension 5	999	958	1,058	3,296
Dimension 6–8	2,084	3,846	1,369	3,080
Heterogeneity	30,579	23,727	21,588	—
Percent decomposition				
Dimension 1	74.24%	74.63%	75.46%	76.31%
Dimension 2	14.35%	14.75%	14.58%	15.65%
Dimension 3	6.06%	5.89%	6.03%	6.19%
Dimension 4	0.32%	0.47%	0.34%	0.90%
Dimension 5	0.15%	0.14%	0.16%	0.49%
Dimension 6–8	0.31%	0.57%	0.20%	0.46%
Heterogeneity	4.57%	3.54%	3.22%	—

TABLE 8.2
Influence Measures from the RC(3) Complete Homogeneity Model for Occupations

| | Current Labor Force Position | | | | | | | | |
| | Unemployed | | | | Part-time employed | | | Full-time employed | |
Labor market experience	NILF	New/ rentrants	Quits/ losses	Layoffs	No full-time	Economic reasons	Voluntary	Low income	Adequate
Nonworkers									
Not looking	**2.39**	0.90	−0.43	−0.50	−0.97	−1.02	0.34	−0.32	−.039
Looking 15+ weeks	0.52	1.60	1.57	0.93	−0.37	−0.57	**−1.92**	−1.09	−0.68
Looking 1–14 weeks	1.34	1.19	0.37	0.11	−0.35	−0.69	−0.43	−0.63	−0.91
Part year–part time									
Not looking	1.25	−0.13	−1.31	−0.94	0.09	−0.28	1.71	0.36	−0.76
Looking 15+ weeks in 1 stretch	−0.33	0.32	0.11	0.11	0.74	0.17	0.18	−0.16	−1.14
Looking 15+ weeks in 2+ stretches	−0.41	0.42	0.34	0.26	0.67	0.15	−0.13	−0.26	−1.04
Looking 1–14 weeks in 1 stretch	0.11	−0.09	−0.62	−0.39	0.57	0.11	0.99	0.19	−0.88
Looking 1–14 weeks in 2 stretches	−0.10	−0.00	−0.33	−0.19	0.57	0.15	0.64	0.08	−0.82
Part year–full time									
Not looking	1.25	−0.05	−0.40	−0.39	−0.54	−0.54	0.11	0.10	0.97
Looking 15+ weeks in 1 stretch	−0.33	0.38	1.09	0.72	−0.37	−0.05	−1.49	−0.43	0.60
Looking 15+ weeks in 2+ stretches	−0.41	0.47	1.07	0.72	−0.13	0.00	−1.36	−0.46	0.22
Looking 1–14 weeks in 1 stretch	0.11	−0.05	0.62	0.41	−0.50	0.00	−0.99	−0.12	0.99
Looking 1–14 weeks in 2 stretches	−0.10	−0.09	0.51	0.36	−0.22	0.11	−0.75	−0.08	0.67
Full year									
Voluntary part time	0.38	−1.55	**−2.22**	−1.42	0.24	0.30	**2.72**	1.22	0.33
Involuntary part time	**−4.00**	**−1.95**	−0.11	0.33	**2.09**	**1.90**	0.62	0.93	0.19
Full time	−0.50	−1.37	−0.27	−0.13	−0.98	0.26	−0.23	0.67	**2.64**

Note: Entries in bold are greater than two standard deviations from the mean.

features of the labor market that stand out. First, those involuntarily part-time employed for the full year prior to the survey are likely to remain part-time employed at the time of the survey for two reasons, because no full time jobs are available and because of economic reasons. Second, those unemployed because they are new entrants and reentrants to the labor market are not likely to have come from the pool of workers who were involuntarily part-time employed for the full year prior to the survey. Finally, those voluntarily part-time employed for the full year prior to the survey are not likely to incur a job loss or to quit a job 1 year hence.

With these additional strong associations, the results here are still similar to what we found earlier in examining race–sex groups. That is, these results show that the structure of the market experience and labor force outcome association is still dominated by (1) those who remain out of the labor force, in voluntary part-time employment, or adequate full-time employment states over the course of a year, (2) those who have a high propensity for not taking part-time work, and (3) those who have a high propensity for settling for part-time work but who would rather have full-time employment.

Equalizing Market Experiences across Occupations

In this section we present labor-force distributions that would have occurred had labor market experiences been orthogonal to occupational locations over time. As we did in Chapter 7, these counterfactual distributions are obtained using the marginal CG purging method described in Clogg and Eliason (1988). In calculating these counterfactual results, we use the *1996 upper nonmanual occupation distribution* as a standard. We chose this standard because upper nonmanual occupations have the most favorable distribution of labor force outcomes from 1983 to 1996.

Figure 8.10 presents the inequality measures for the observed and purged distributions. (See the previous chapter for details on this measure.) Equalizing prior labor market experiences reduces inequality by 24% over the observed distribution, and this reduction is relatively stable from 1983 to 1996.

Adequate Full-Time Employment. Figure 8.11 gives the purged odds of adequate full time employment. Comparing these odds to those in Fig. 8.1 shows how orthogonalizing the experience–occupation relation would change the odds of adequate full-time employment. First note that the baseline increases from an observed odds of 14.2 to a purged odds of 22.7. Thus, equalizing market experiences across occupations in this manner increases the baseline chances of adequate full time employment by almost 60% above the observed odds. Additionally, it appears that administrative support occupations would gain some ground relative to upper nonmanual occupations.

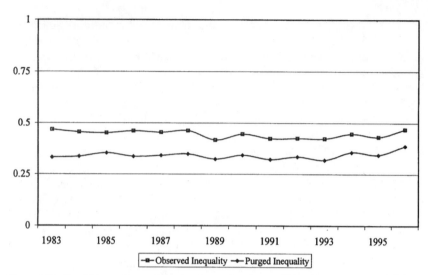

Figure 8.10. Overall inequality measures for observed and purged labor-force outcome distributions (overall observed inequality = 0.45; overall purged inequality = 0.34).

Comparing the observed and counterfactual purged odds, it appears that differences in the odds of adequate full-time employment would be smaller if labor-market experiences were equalized. If labor market experiences were equalized across occupations, occupational inequality in adequate full-time

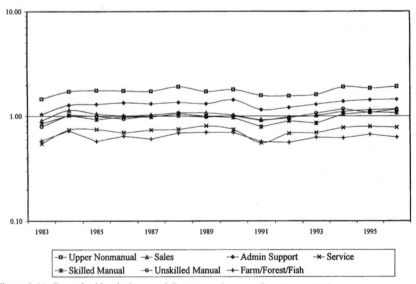

Figure 8.11. Purged odds of adequate full-time employment (baseline = 22.7).

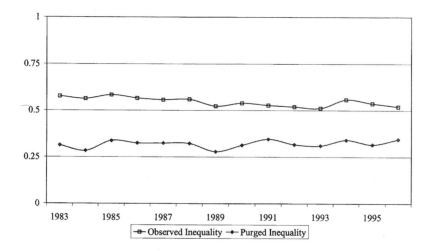

Figure 8.12. Inequality measures for adequate full-time employment (overall observed inequality = 0.55; overall purged inequality = 0.32).

employment would drop by 42%, as suggested by the inequality measures shown in Fig. 8.12. This result suggests that there is considerable inequality across occupations in access to stable, year-round employment and that this inequality has been quite durable over time.

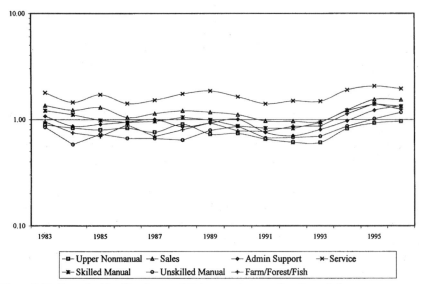

Figure 8.13. Purged odds of low-income full-time employment (baseline = 2.6).

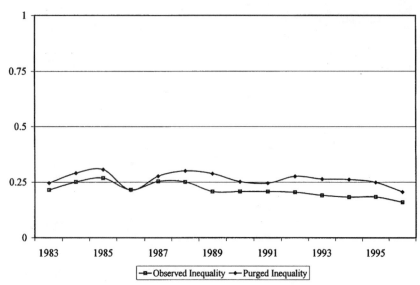

Figure 8.14. Inequality measures for low-income full-time employment (overall observed inequality = 0.22; overall purged inequality = 0.26).

Low Income Full-Time Employment. The story for low income full-time employment differs greatly from this. Fig. 8.13 shows the purged odds. The odds of low-income full-time employment would increase 37% if prior labor market experiences were equalized. Service occupations would be most affected. The inequality measures in Fig. 8.14 indicate that orthogonalizing the experience–occupation relation does not have a large impact on inequality among occupations in the odds of low income full-time employment. Indeed, at most time points Fig. 8.14 shows that the purged odds exhibit higher levels of inequality than the observed odds.

Voluntary Part-Time Employment. Recall that the observed odds in Fig. 8.3 suggested two distinct classes for voluntary part-time employment: Those in white-collar occupations had consistently higher odds of voluntary part-time employment, while those in blue collar occupations had consistently lower odds of voluntary part-time employment. Had market experiences been independent of occupations, this split would no longer be as strongly delineated, as indicated by Fig. 8.15. Although the baseline odds would change little, with 2.3 for the observed and 2.6 for the purged odds, the patterns of inequality would change considerably. Figure 8.15 shows that, under the purged odds, upper nonmanual and administrative support occupations would have the highest odds, with sales and service occupations dropping noticeably below the top two for most of the time period. The odds among blue-collar occupations make up the bottom tier.

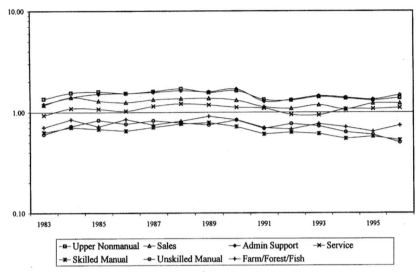

Figure 8.15. Purged odds of voluntary part-time employment (baseline = 2.6).

Figure 8.16 shows further that inequality in the odds of voluntary part-time employment would be reduced had market experiences been independent of occupations. The overall inequality measures show a decline by a factor of [0.33/0.58] = 0.57 of the base. Comparison of the trends also shows that inequality would be reduced throughout the years 1983 to 1996.

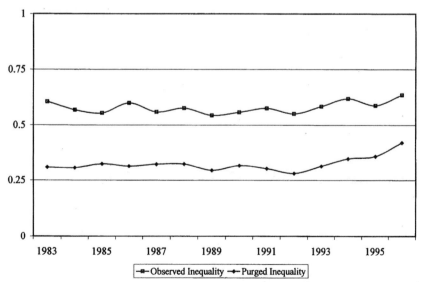

Figure 8.16. Inequality measures for voluntary part-time employment (overall observed inequality = 0.58; overall purged inequality = 0.33).

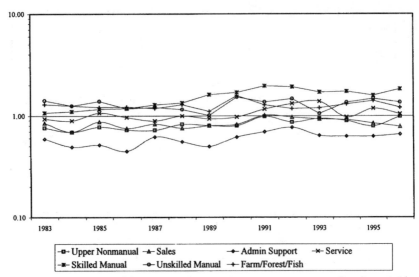

Figure 8.17. Purged odds of part-time employment due to economic reasons (baseline = 0.6).

Part-Time Employed for Economic Reasons. As in the comparison of race–sex groups in Chapter 7, a comparison of Figs. 8.17 and 8.4 shows that there is little difference between the observed and purged odds of being part-time employed due to economic reasons. Not only do the baseline odds remain the same, a comparison of the trends shows remarkable similarities. Figure 8.18 further shows that there is practically no difference in the inequal-

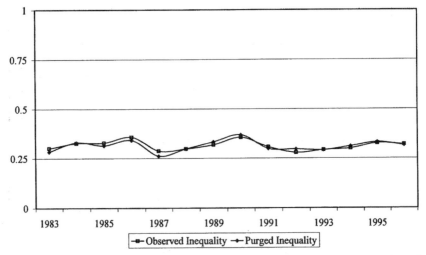

Figure 8.18. Inequality measures for part-time employment due to economic reasons (overall observed inequality = 0.31; overall purged inequality = 0.31).

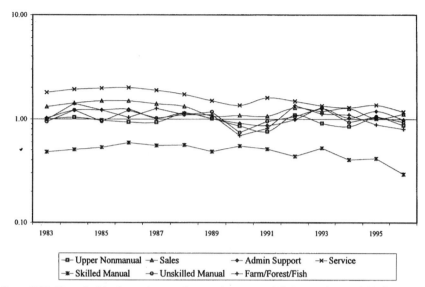

Figure 8.19. Purged odds of part-time employment because no full-time work is available (baseline = 0.4).

ity among occupations for the purged and observed cases. All of this indicates that rendering market experiences independent of occupations does little to change the odds of part-time employment due to economic reasons.

Part-Time Employed—No Full-Time Available. Figure 8.19 shows that the baseline purged odds of part-time employment because no full time employment is available, 0.4, does not differ all that much from the observed baseline odds, 0.5. The trend for the purged odds is somewhat similar to that for the observed, though the across-occupation variability in purged odds is more tightly packed. Figure 8.20 provides the corresponding inequality measures. If market experiences were independent of occupations, inequality in the odds of part-time employment for lack of full time jobs would have been reduced throughout the time period.

Unemployed—Layoffs. Comparing Figs. 8.21 and 8.6 indicates that the baselines for the observed and purged odds of being unemployed due to layoff are the same. Although the odds of being laid off would still fall along blue/white collar lines had market experience been independent of occupations, Fig. 8.21 shows that that gap would diminish somewhat.

Figure 8.22 gives a more precise view of the lessening inequality in the odds of being laid off had market experiences been independent of occupations. As indicated by both the overall inequality kappas and the trends, inequality among occupations in the odds of being laid off clearly would be reduced were experiences independent of occupations.

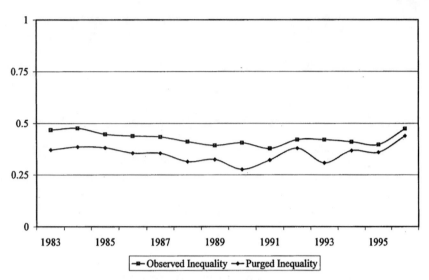

Figure 8.20. Inequality measures for part-time employment because no full-time work is available (overall observed inequality = 0.43; overall purged inequality = 0.35).

Unemployed—Quits and Losses.   Orthogonalizing the experience–occupation relation has a more noticeable impact on being unemployed due to quits and losses than in the case of unemployment due to layoffs. A comparison of Figs. 8.23 and 8.7 shows that, although the baselines differ very little

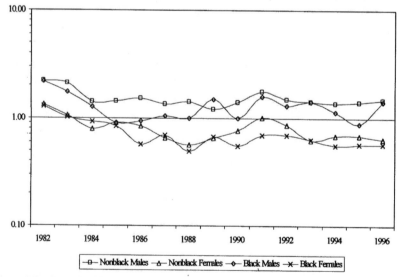

Figure 8.21. Purged odds of unemployment due to layoffs (baseline = 0.2).

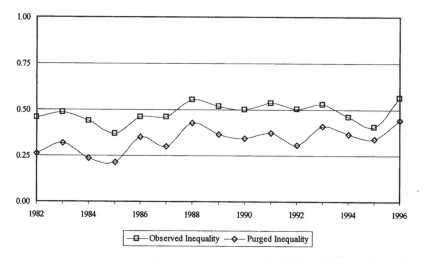

Figure 8.22. Inequality measures for unemployment due to layoffs (overall observed inequality = 0.81; overall purged inequality = 0.67).

between the observed and purged odds, the blue/white collar gap in the odds of this type of unemployment would mostly disappear if market experiences were independent of occupations. The inequality measures in Fig. 8.24 show further that the overall inequality would be reduced by a half, with a greater reduction observed at the beginning of the time period than at the end.

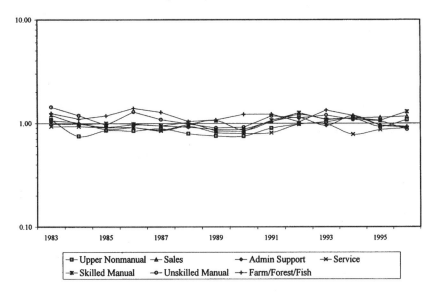

Figure 8.23. Purged odds of unemployment due to quits and job losses (baseline = 0.7).

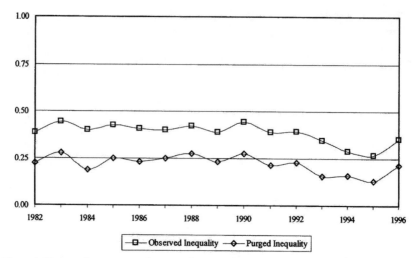

Figure 8.24. Inquality measures for unemployment due to quits and job losses (overally observed inequality = 0.26; overall purged inequality = 0.13).

Unemployed—New Entrants and Reentrants. Orthogonalizing the experience/occupation relation has little effect on the odds of being an unemployed new entrant or reentrant into the market. A comparison of Figs. 8.25 and 8.8 shows that the baseline odds and the general pattern change little over time. An exception to this is the relatively low drops for unskilled manual

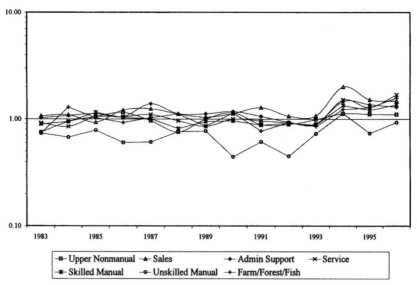

Figure 8.25. Purged odds of new entrant and reentrant unemployment (baseline = 0.3).

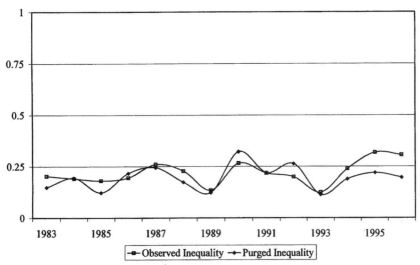

Figure 8.26. Inequality measures for new entrant and reentrant unemployment (overall observed inequality = 0.23; overall purged inequality - 0.20).

occupations in 1990 and 1992. Figure 8.26 indicates that the inequality measures show very little difference overall, and in the trend as well.

Not in the Labor Force. Although a comparison of the baseline odds of not being in the labor force, as shown in Figs. 8.27 and 8.9, and the inequality

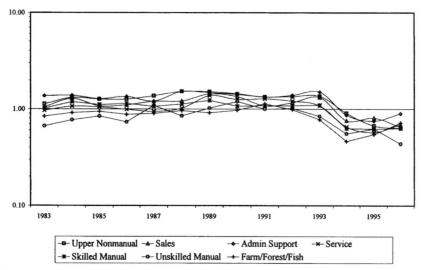

Figure 8.27. Purged odds of not being in the labor force (baseline = 0.6).

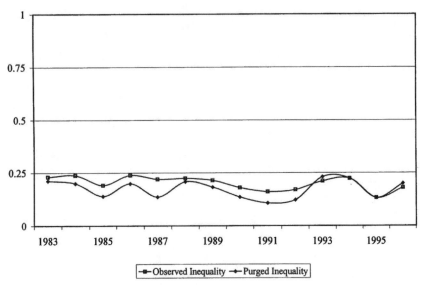

Figure 8.28. Inequality measures for not being in the labor force (overall observed inequality = 0.20; overall purged inequality = 0.18).

measures shown in Fig. 28, indicates little difference between purged and observed odds, the final point in the trend, 1996, does indicate a diminished blue/white collar gap.

DISCUSSION AND CONCLUSIONS

Our empirical investigation has revealed the following key findings. There is very little difference in the trend in the relationship between labor force outcomes and the market experiences of respondents during the prior year. Temporal change accounted for only around 3% of the overall total association between experience and outcomes.

Second, there are few across-occupation differences in the relationship between labor force outcomes and the market experience of respondents during the prior year. Across-occupation differences in the experience–outcome relationship accounted for 3.5% of the overall total association.

Third, the relatively stationary and stable experience–outcome relationship is largely governed by a set of four pockets of strong positive attractions and three pockets of strong negative repulsions. The positive attractions consist of (1) a strong tendency to retain adequate full-time employment, (2) a strong tendency to remain voluntarily part-time employed, (3) a strong tendency for those out of the labor force and not looking for work to remain out of the labor force, and (4) a strong tendency for those who are involuntarily part-time employed to remain so. The negative attractions include (1) a tendency for nonworkers who have looked for 15 or more weeks to not accept part-time employment, (2) the tendency for the voluntarily part-time employed to not quit or lose their jobs, and (3) a tendency for those who are involuntarily part-time employed to not drop out of the labor force.

When prior labor market experiences are equalized across occupations, inequalities in labor force outcomes are reduced by 24%. The odds of adequately compensated full-time employment increases by 60%. Because the purged odds are produced by standardizing all occupations to the experiences of upper nonmanual occupations in 1996, these results suggest that one positive policy intervention would be to equalize prior labor market experiences across groups. This could be done by (for example) providing tax incentives for the creation of stable, full-time, year-round jobs in specific occupations.

Our results also indicate that there would be unintended negative consequences if such standardization were achieved. The odds of low income full-time employment would increase. Such a policy would do little to change the baseline odds of most types of part-time employment or unemployment. But inequality in voluntary part-time employment and unemployment created by layoffs, quits and job loss would be reduced considerably. These are the main forms of occupational inequality discussed by students of economic restructuring (Bluestone & Harrison, 1982; Gordon, 1996).

Empirically this analysis points to one of the major differences between the U.S. labor market and many of its European counterparts. Western European governments help considerably in the creation of adequately paid, year-round, full time work at the expense of lesser forms of labor force attachment. Equalizing prior labor market experiences in the U.S. context would look a lot like the extensive interventions in the labor market that occur in Western Europe. However, Western European labor markets also have very high rates of unemployment and labor force nonparticipation. This may be a trade-off for the interventions that are necessary to produce better-paid, full-time jobs and stable work commitments by employers.

To combat this trade-off, we suggest that policymakers think about tax incentives for spurring full-time, year-round employment. Rather than (implicitly) rewarding employers for creating jobs of any kind, many of them part time, unstable, and of uncertain quality, we think it would be better to provide special sets of incentives to produce year-round full-time jobs. These incentives would combat the current incentives to create large numbers of marginal jobs in favor of producing smaller (but still considerable) numbers of good jobs. The current Western European system does not reward employers for job creation and barely penalizes them for remaining stagnant. A U.S.

based system, more in tune with our political situation and values, would reward employers for creating better jobs and still permit the creation of all types of jobs. Hopefully the incentives would be good enough that employers on the verge of creating stable, full-time jobs for workers would go ahead and do so.

Our results also have some implications for debates about "welfare to work" policies as components of welfare reform. Our results suggest that there is considerable segmentation of the labor market and that the segments display a great deal of stability and inertia. There are considerable barriers to steady and permanent movements from nonparticipation to participation in the labor force, and from various types of part-time work to stable, full-time, year-round work. This would not be a problem if jobs short of full-time, year-round work paid enough to justify leaving welfare, but most do not (see Jencks, 1992). The movement of welfare recipients onto the work roles into jobs that will provide enough attachment to the labor force, enough hours of work, enough steadiness of demand for their labor, and enough pay to consistently remain above the poverty line is a considerable and nontrivial undertaking that will require much more effort than simply removing people from the welfare roles.

Finally, while we found pockets of considerable rigidity and segmentation, there were also (correspondingly) areas of relatively free movement into and out of different labor market positions. The major problem for social policy, to which our analysis attests, is that relatively little of that fluidity involves movement into full-time, year-round work. Instead, labor force participants seem to hop between leaving the labor force, voluntary part-time employment, unemployment, and part-time employment combined with further searches looking for a full-time job. We think it will take considerably more research to isolate how much of this fluidity is voluntary and how much of it is part of a prolonged search for adequate employment at a satisfactory level of earnings.

NOTES

1. Any more detailed occupational classifications would lead to sparce cell frequencies. We begin our analysis in this chapter in 1983 rather than 1982 because the 1980 Census Occupation Codes were not in use in the March 1982 CPS.

2. These odds are calculated in the same manner as the odds in Chapter 7 for race and gender groups. See Chapter 7 for details.

9

Toward a More Complete Understanding of Labor Markets and Stratification

The purpose of this book was to introduce to a broad constituency of readers some of the basic issues that Clifford Clogg and his colleagues dealt with in their studies of the labor force. Before we examine what we think are the policy implications of this analysis, it is worth summarizing briefly what the analysis has told us so far:

1. *Unemployment, by itself, does not adequately describe the multiple dimensions of marginal involvement in the labor market.* The original reservations about using unemployment rates as the single indicator of how the labor market is performing seem to be right on target and (if anything) the conclusions reached by researchers in the 1930s are more true today than ever before. Researchers in social stratification and policy analysis have long suspected that within the omnibus categories of "employment" and "unemployment" there were people in distinctly different positions in the labor market whose access to work opportunities varied considerably.

The analysis we presented here was designed to drive home one of the central substantive points of Clogg's research on the labor force. *Access to full-time, year-round work cannot be assumed and a great deal of structural inequality in labor markets can be explained by who has access to full-time, year-round work and who does not.* Clogg's work suggests that, at most, about 60% of the labor force has access to full-time, year-round jobs. The remaining 40% of the population is either marginally employed, unemployed and looking for work, or out of the labor force altogether. These numbers hold primarily for nonblack males.

The percentages of year-round adequately employed for other groups fall substantially behind those for nonblack males. This was true in the early 1980s and has remained true through 1996.

In contrast to most students of social stratification, Clogg and his colleagues focused on labor market marginality as a central determinant of locations in the stratification system. Much research in this area (including our own, cf. Leicht, Wallace, & Grant, 1993; Eliason, 1995a) focuses on stratification among those who have access to full-time jobs. While this is certainly an important and salient aspect of the production of social inequality, Clogg's work served to highlight that a considerable amount of social inequality was missed by studying only those who were employed full time. If the only thing one took away from Clogg's analysis was the idea that social stratification should address inequality among the marginally employed in addition to the fully employed, that would lead to considerable advancements in our understanding of how labor markets work in the United States.

2. *Labor market segmentation can be characterized by access to full-time, year-round work.* Along with defining labor market segmentation in the traditional fashion of industry or firm location, Clogg's analysis presents yet another important dimension along which markets are structured. A major delineation in the labor market exists between those who do have access to full-time, year-round work and those who do not. There is a considerable structural barrier between these groups that is difficult to penetrate from any of the marginal labor market positions we examine. A substantial number of labor force entrants have difficulty gaining access to full-time work, and those who have access to it rarely (if ever) occupy a different labor market niche. While full-time, year-round work systematically varies by occupational location (an issue we address below), this type of work is available in a variety of industries, firms, and work settings. Those who have access to this work are definitely advantaged in the labor market relative to those who do not. We think that descriptions of labor market segmentation, and its effect on market attainments, that do not include this labor force dimension are at best incomplete and at worst misleading with respect to understanding the unequal distributions of market outcomes.

3. *Labor force measurement adds considerably to the explanatory power of traditional stratification models predicting socioeconomic status and earnings.* The contributions of labor force measures to explaining these important stratification outcomes are roughly equivalent to the contributions of age and education. Further, measures of prior labor force experience provide an important intervening mechanism between traditional measures of ascription and achievement and desirable stratification outcomes.

The role of labor force experience as an intervening mechanism is most apparent in analyses of women and underrepresented groups compared with white men. Specifically, the addition of prior labor force experience helps to highlight a key mechanism involved in the production of race and gender inequality in labor market outcomes. *Age and education aid the labor market prospects of marginalized groups by moving incumbents between labor force categories*, specifically in the direction of full-time, year-round work. *Age and education help nonmarginal groups improve their stratification prospects by improving their position within labor-force categories*, rather than between them. Taken together, these results suggest that white men, in particular, are privileged when it comes to access to full-time work, regardless of their achieved or ascribed characteristics. Women and other underrepresented groups move in the stratification system to the extent that their other achieved and ascribed characteristics move them in the direction of attaining full-time work. These results highlight the critical role played by prior work experiences, a theme that we will return to later in this chapter.

4. *There are considerable over-time shifts in labor force categories, and most of these shifts are away from adequate employment.* The shifts across the decades of the 1970s and 1980s, in particular, suggest that there was an overall 10% decline in the number of adequately employed workers and a corresponding increase in marginal labor-force categories, especially the educationally mismatched. Further, almost all of the movement across labor-force categories during recession years occurred among those in the underemployed categories of the labor force. In no case do we find systematic movements out of adequate employment toward marginal labor force categories during recessions, only to see the ranks of full-time workers expand during economic recoveries. Instead, *it appears that there has been a long-term, secular decline in access to adequate employment.* This shift, and the corresponding increases in underemployment, is the central over-time trend in labor force opportunities that we uncover. If nothing else, this reinforces our point that the study of labor market marginality is more important now than it has ever been.

5. *Almost all of the differences across race, gender, and occupation in the relationship between prior labor-market experiences and current labor force outcomes result from systematic differences in the distribution of experiences*, rather than differences in how prior experiences translate into current outcomes. Equalizing prior labor market experiences across groups produces considerable reductions in inequality across labor market positions; declines range from 20–30% to almost 60%, depending on the labor-force outcome we examine.

In our observed distributions of labor-force outcomes over time, nonblack men tend to dominate full-time, year-round work. White women dominate the voluntary part-time employment category. Black women dominate short-term unemployment categories and marginal employment categories, and black men occupy niches with long unemployment spells, long job searches, and little in the way of part-time employment prior to full-time employment. Further, there is considerable evidence of growing inequality among blacks in labor-force outcomes. Black men, in particular, increasingly are "all the way in" or "all the way out" of the labor market, increasing their numbers in full-time work categories but also increasing their numbers in unemployment categories.

The analysis of labor-force categories across occupations produced information that corresponds to other prior descriptions of the creation of a postindustrial work force (see Harrison & Bluestone, 1988; Rifkin, 1995; Moore, 1996). Specifically, we found almost no evidence of "managerial downsizing" in the 1990s. Instead, there were marginal shifts toward jobs in service occupations. Service occupations disproportionately have incumbents in low-income full-time jobs, various kinds of part-time jobs, little in the way of involuntary part-time work, and reduced risks of layoffs. Manual occupations (relative to service occupations) produce more full-time work, produce relatively little part-time work, and drastically increase the chances of layoffs. Overall, these results suggest that compositional shifts in the labor market have moved employment away from occupations that traditionally provide access to full-time work and toward occupations where full-time work is either poorly compensated or not available. These results provide more systematic support for the work of other researchers (see Bluestone & Harrison, 1982; Wallace & Rothschild, 1988; Perrucci & Targ, 1993; see also Milkman, 1997).

POTENTIAL IMPROVEMENTS AND A FUTURE RESEARCH AGENDA

Obviously, we think that Clogg's contributions to the study of labor markets and social stratification have been important. Our extensions of his analyses in Chapters 7 and 8 suggest some new insights that we think are possible through the extension of Clogg's basic research. This does not mean that Clogg's research could not be improved and expanded. We outline several fruitful extensions of Cliff's work here that go beyond our preliminary attempts to build on the work he started.

The most obvious extension of this research would be *to begin to study labor force careers as a significant labor market dimension.* Most of Clogg's analysis and our own uses the yearly March Current Population Surveys. These surveys have many advantages including the representativeness of the samples, coverage of basic demographic information, and the ability to measure labor force activity across an entire year. What the CPS lacks is any ability to trace systematic changes over the work careers of individuals and the empirical regularities these changes create. Such analyses either require longitudinal data of longer duration (for example, the Panel Study of Income Dynamics or the NLSY), or the construction of synthetic cohorts using people at different stages of their work careers. Both kinds of work would allow us to chart systematic movements between underemployment, unemployment, and full-time employment over longer stretches of time. Even synthetic cohort portrayals of such trends would be a step above where our analysis was able to take us at this point.

What would such a synthetic cohort analysis look like? One possibility would be to take people with specific mixes of demographic and occupational characteristics and chart the typical probability of year-to-year movements across labor-force categories for different ages of workers at the same time points. This technique would mirror the synthetic cohort analyses of career lines conducted by Otto, Spenner, and Call (1980a, b; see also Spenner, 1982). From studies of poverty and labor markets (see Jencks, 1992; Wilson, 1996; Hauan et al., 2000) we know that the poor vacillate repeatedly between bouts of marginal employment, unemployment, "off the books" employment, and idleness. We have a much more limited impression of the basic demographics of labor-force positioning across the entire population. Even research using synthetic cohorts to address the issue of labor force careers would provide a set of benchmarked profiles for the evaluation of the activities of groups that policy analysts spend a good deal of time discussing (the poor, welfare recipients, underrepresented groups, the aged, and so on).

A second useful extension would be to *monitor changes in labor force characteristics across selected sets of detailed occupational groups*. Most research in social stratification has concluded that there is growing inequality within occupations, firms, and even jobs (see Petersen & Morgan, 1995). There are serious problems of comparability in the occupation codes used in different Census years, and these problems are mirrored in the CPS. However, we could take specific occupational groups (such as managers) and conduct more detailed analyses of labor force activity within broad occupational groups. A more detailed analysis of managers and other white-collar occupations seems warranted in light of anecdotal evidence that there has been considerable downsizing and elimination of middle management and white-collar jobs during the 1990s (see Rifkin, 1996; Moore, 1996; Dunkerley, 1997; Leicht, 1998; but see Gordon, 1996). Several interesting future research questions would fall under this heading; has there really been a noticeable decrease in the availability of full-time managerial work? Is the appearance of increased administrative ratios discussed by some scholars (most notably Gordon, 1996) caused by layoffs of nonsupervisory employees, corresponding increases in full-time supervisory employment, or declines in both categories occurring at different rates? There seem to be more than a trivial number of personal stories about the increasing hardships of employees who were once exempt from economic restructuring (*New York Times*, 1996). What is not clear at this point is whether these stories reflect general demographic trends or stories of unfortunate individuals experiencing bad luck. More systematic analyses of labor force trends within occupational groups would bring systematic evidence to bear on this issue.

Third, to expand the purview of the LUF as a mechanism for studying labor market inequalities, *the LUF indicator needs a "top."* Right now the most privileged category in the LUF that has driven our analysis is full-time, year-round work with earnings at one and a half times the poverty line. We have shown that there is considerable labor market stratification produced by simply knowing who is in this category, who can stay there, and who cannot manage to get there.

But to further integrate studies of the labor force with other traditions for studying labor markets, we think the LUF could use some new categories that more specifically described privileged locations in the labor market. Other research focuses on three distinct dimensions of labor market privilege that affect workers' welfare and labor-market experiences: *access to skilled jobs, access to fringe benefits,* and *access to internal labor markets.* Labor force entrants can have a full-time job that meets the criteria of the most privileged labor utilization category, but have a job that varies widely in the use and creation of job skills, access to fringe benefits (which constitute between 20 and 60% of the average employment bill), and access to established career ladders. In spite of our claim that limiting research to full-time workers misses significant dimensions of labor market inequality and social stratification, prior research has spent much time discussing these distinctive dimensions of labor markets and work careers as producers of social stratification. We think that future efforts could incorporate these dimensions at the top of standard treatments of the labor force and analyze social inequality produced by the labor market across an entire range of relative deprivation and privilege. Such an analysis would truly integrate both the marginally employed and the elite of the U.S. labor market, providing insights into the barriers to entry and distinctive achieved and ascribed characteristics that sort people into different labor market segments.

Fourth, *future research should revisit the concept of educational mismatch and attempt to deal with credential inflation as an issue in the study of labor market inequalities.* We were unable to continue usage of the original educational mismatch measure that Clogg developed and we suspect that the old categorization may have lost its utility in our analysis extending into the late 1980s and 1990s. However, this does not lessen the need to study "educational underemployment" and the changing use of credentials by employers and potential employees. The old mismatch measure was unable to chart secular increases in credentials across occupations or account for the reasons for these changes. Future research on labor market inequalities should re-visit this issue and try to face up to the changing ways that educational credentials are being used in the labor market (see Stolzenberg, 1994; Leicht, Hogan, & Wendt, 2000). One obvious way to do this would be to chart the inflation in educational credentials across the same occupational groups over time, creating different measures of educational mismatch based on the relative standing of different educational credentials at different time points. Then an analysis of labor-force trends over the past 30 years could ask some interesting questions such as, is the demand for college educated labor driven by the creation of jobs that traditionally require college credentials to perform, or is the demand for college educated labor driven by employers who replace those without college credentials with college graduates because college graduates are available at lower prices? Such questions have received very little attention since the publication of Ivar Berg's book *Education and Jobs: The Great Training Robbery* (New York: Praeger, 1970).

Finally, *future research should pay more attention to the growing flexibility in work routines produced by the debureaucratized workplace.* Numerous authors have discussed the changes in the workplace in the 1990s caused by increasing globalization and the availability of flexible automation (see Florida & Kinney, 1990; Moore, 1996; Dunkerley, 1996; Leicht, 1998). The growing use of temporary workers, subcontracting and outsourcing, flatter organizational hierarchies, and the general downsizing of the permanent work force must have some cumulative effect on the labor force. These changes produce some challenges that Clogg could not have foreseen that will require new subtleties in measurement and interpretation. For example, compensation may now come in the form of stock options, single bonus payments, and other schemes that limit year to year earnings while maximizing long term cumulative economic gains. Consultants and other "outsourced" workers may have patterns of work activity that look much more like self-employment than regular contact with the labor market as an employee. Occupational categories may change in their composition and meaning in unpredictable ways. Entire subsegments of the population may be left out of the race for gainful employment altogether. These issues, and others, will present challenges to anyone interested in explaining labor market inequality.

POLICY RECOMMENDATIONS THAT RESULT
FROM STUDIES OF THE LABOR FORCE

The policies that derive from the analysis presented here depend critically on the values that policy analysts bring to their interpretation of our results. When we think about the policy implications of this work, we suspect we share the values that most social scientists have. To make sure of this, we will summarize these commitments in a few sentences and then discuss the policies that follow from our analysis in light of these commitments.

The first commitment many people who study social inequality have is to *meritocracy and equity*. Individual effort and investment in one's future should produce outcomes commensurate to that investment and effort. The production of systematic differences in labor market outcomes between people with similar investments is a situation that undermines the legitimacy of the economic system and leads to the perception that some groups have advantages in the labor market that they do not deserve.

The most obvious and egregious examples of this are cases of working poverty (situations where people work at full-time jobs but make wages that fail to push them above the poverty line) and cases of educational mismatch (where people who are consistently overqualified for certain positions end up in those positions repeatedly). If both situations are widely perceived then these reduce normative commitments to the functioning of the labor market and calls for returns to meritocracy and equity.[1]

The second commitment that most researchers in social stratification have to some degree is a commitment to *reducing inequality*. Few of us doubt that some inequality is necessary to send appropriate market signals, to get certain types of people to do certain things, and to otherwise motivate people to participate in the labor market at some level. What most of us doubt is that the types of earnings and income inequality produced in the United States are necessary to send these labor market signals. Hence most debates are about the amount of inequality that is necessary rather than whether there should be inequality at all (an important distinction).

Finally, most students of stratification and labor markets are committed to the concept of *freedom of choice*. People should be able to pursue their own idiosyncratic economic objectives without the hindrance of structural constraints, gender and racial discrimination, or accidents of birth getting in their path. People should be able to make different types of commitments to the labor market at different stages of their lives without incurring penalties that hopelessly foreclose future choices (this is not to say that there should be no trade-offs at all). Labor markets that narrowly constrain individual choice without providing clear and coherent explanations for it risk losing the legitimacy of the population on whose work most economic productivity depends.

Now that we have said all this, what does our analysis suggest could be done to reduce labor market inequalities, especially those that are not the product of conscious individual choices? First, *we need to spend more research effort figuring out how much of the labor force activity we see is a product of labor market constraints or conscious choices.* As some researchers have pointed out, eventually all decision-making can be viewed as a product of constraints (see Coleman, 1990; Reskin & Roos, 1990). But that is not necessarily what we mean in this context.

If one looks at the analysis we present here, several things strike us as worth more research effort in an attempt to decide what policy interventions would be appropriate. There are large numbers of people over time who are unemployed or who have marginal attachment to the labor market. A certain percentage of this marginal attachment is voluntary, decisions made in light of other commitments and choices. But much of this commitment is the result of artificial constraints on choice that are imposed by the structure of labor markets.

Take our full-time, adequately employed labor force category for example. Approximately 60% of the population at any one time occupies this labor force category. But (unlike the underemployment categories that imply that people are seeking greater labor-market attachment and rewards) how many people really want to be there? There is growing evidence that the hours of work among full-time workers is increasing (see Shor, 1992, 1998). Further, in the United States, full-time work is often necessary to secure the earnings and fringe benefits necessary for most people to make ends meet. There are very few half-time jobs that really are "half" of a full-time job; the work is often "two-thirds" while the compensation and benefits are more like "one-quarter" or less. Hence, people are trapped in full-time employment who do not want to be there at the same time as others desperately seek full-time employment for the earnings and benefits that only it will provide.

One tentative solution to this problem is to provide tax incentives to employers to provide better pay and fringe benefits for part-time workers and to permit full-time workers to "scale down" their jobs through job sharing plans. Employers have a vested interest in producing jobs in full-time equivalent chunks (Schor, 1992). Such jobs produce workers who are more marginally productive than a larger number of people each making a half-time commitment (or varying their work hours). Yet the creation of full-time jobs and the expansion of work hours required in them beyond the 40-hour workweek not only puts a strain on full-time workers (who must balance their commitments to work against families and other obligations), it prevents others who are looking for greater labor market attachment from attaining rewards. The use (and overuse) of full-time workers is a classic "public goods" problem that will not go away without government intervention. Individually, each employer has an incentive to work his or her full-time staffs harder and harder while employing as few people as he or she can possibly get away with. The only way out of this conundrum is to provide incentives for employers to do otherwise.

A second place where public policy should be focused is in the relative equalization of labor-market experiences across subgroups. Our analysis clearly shows that there are few systematic differences in the ways that prior experience is linked to current positions in the labor market. Apart from conducting more systematic analyses across individual work careers (something we suggested above), the place where there are clear differences across subgroups is in the systematic accumulation of meaningful labor-market experiences that translate into relatively privileged positions in the labor market. Policies that attempt to match people to jobs, get new labor market entrants into the labor market, and that move marginalized labor market groups toward full participation will go a long way toward reducing systematic differences across race and gender in access to privileged labor market positions.

A focus on improving labor market matching across different levels of prior labor-force experience will do something that is politically useful as well. Most politicians and the polity in general define the "deserving" poor as those who are working and making a normative commitment to contact with the labor market. In spite of considerable evidence that welfare recipients and others who are poor have at least intermittent contact with the labor market (see Jencks, 1992), most people believe that the "undeserving poor" do not work. In this climate, virtually anything that allows more people to visibly demonstrate their commitment to the labor market will increase perceptions that some segments of the poor "deserve" help and should receive it. Further efforts to ensure that those who make commitments to the labor market receive the requisite rewards that they deserve would also inspire those who are discouraged by market discrimination and structural constraints in the current economic climate.

Our final policy recommendation follows from this observation and the relative distribution of people across labor force categories in our analysis. There are obviously considerable numbers of people across all times and places who could be characterized as "working poor." Virtually no one agrees that working poverty is good for the economic and social fabric, yet there is relatively little consensus about what to do about the situation. There are also disagreements about the relative size of the "working poor."

Our analysis suggests that somewhere in the neighborhood of 20% of the population is "working poor" based on either inadequate income or inadequate hours worked in light of preferences for more work. Regardless of where one stands on the debate about the working poor, the "working" should not be "poor." Either tax incentives should be provided to employers who will pay workers wages that lift them above the poverty line, or the government should subsidize the working poor with a preset package of fringe benefits and in-kind subsidies (see Jencks, 1992). But it is difficult to see how we are supposed to enforce normative commitments to work when working (by itself) will not erase poverty.

These final thoughts are only meant to emphasize the critical and nontrivial nature of the analysis of the labor force for students of labor markets and social stratification. But the importance of this analysis extends far beyond that to the real lives that people face everyday in a society with commit-

ments to distributive justice and equal opportunity for all. In the end, analyzing labor markets and shifts in the labor force is important because it can
contribute to this goal.

NOTES

1. We reluctantly use the term *meritocracy* to describe social systems that reward people
based on what they know and have learned rather than whom they know or accidents of birth.

References

Althauser, R. P., & Kalleberg, A. L. 1990. "Identifying Career Lines and Internal Labor Markets within Firms: A Study in the Interrelationships of Theory and Methods," in R. L. Brieger (Ed.), *Social Mobility and Social Structure* (pp. 308-356). London: Cambridge University Press.

Antos, J., Mellow, W., & Triplett, J. 1979. "What Is a Current Equivalent to Unemployment Rates of the Past?" *Monthly Labor Review* 102:36–46.

Bancroft, G. 1958. *The American Labor Force: Its Growth and Changing Composition.* New York: Wiley.

Beck, Horan, & Tolbert.1978. "Stratification in a Dual Economy: A Sectoral Model of Earnings Determination. *American Sociological Review* 43:704–720.

Becker, G. 1975. *Human Capital: A Theoretical and Empirical Analysis with Special Reference to Education.* Chicago: University of Chicago Press.

Becker, M. P. 1989. "Models for the Analysis of Association in Multivariate Contingency Tables." *Journal of the American Statistical Association* 84:1014–1019.

Becker, M. P. 1990. "Maximum Likelihood Estimation of the RC(M) Association Model." *Applied Statistics* 39:152–167.

Becker, M. P., & Clogg, C. C. 1989. "Analysis of Sets of Two-Way Contingency Tables Using Association Models." *Journal of the American Statistical Association* 84:142–151.

Berg, I. 1970. *Education and Jobs: The Great Training Robbery.* New York: Praeger.

Bielby, D., & Bielby, W. 1984. "Work Commitment, Sex Role Attitudes and Women's Employment." *American Sociological Review* 49:234–247.

Bills, D. B. 1988. "Educational Credentials and Hiring Decisions: What Employers Look for in New Employees." *Research in Social Stratification and Mobility* 10:1–35.

Bishop, Y. M. M., Fienberg, S. E., & Holland, P. W. 1975. *Discrete Multivanate Analysis: Theory and Practice.* Cambridge, MA: MIT Press.

Blau, P., & Duncan, O. D. 1967. *The American Occupational Structure.* New York: Wiley.

Bluestone, B., & Harrison, B. 1982. *The Deindustrialization of America.* New York: Basic Books.

Bluestone, B., & Harrison, B. 1986. The Great American Job Machine: The Proliferation of Low-Wage Employment in the U.S. Economy. Washington, DC: U.S. Congress Joint Economic Committee.

Bowen, W. G., & Finegan, T. A. 1969. *The Economics of Labor Force Participation.* Princeton, NJ: Princeton University Press.

Bureau of Labor Statistics. 1987. Linking Employment Problems to Economic Status, Bulletin 2282. Washington, DC: Government Printing Office.

Clark. R. L., & Menefee, J. A. 1980. "Economic Responses to Demographic Fluctuations," in *Special Study on Economic Change, Vol. 1. Human Resources and Demographics: Charactenstics of People and Policy* (pp. 1–31). U.S. Congress, Joint Economic Committee.

Carnoy, M. 1994. *Faded Dreams: The Politics and Economics of Race in America.* London: Cambridge University Press.

Clogg. C. C. 1978. "Adjustment of Rates Using Multiplicative Models." *Demography* 15:523–539.

Clogg. C. C. 1979. *Measuring Underemployment: Demographic Indicators for the United States.* New York: Academic Press.

Clogg. C. C. 1980. "Characterizing the Class Organization of Labor Market Opportunity: A Modified Latent Structure Approach." *Sociological Methods and Research* 8:243–272.

Clogg. C. C. 1981. "Latent Structure Models of Mobility." *American Journal of Sociology* 86:836–868.

Clogg. C. C. 1982a. "Cohort Analysis of Recent Trends in Labor Force Participation." *Demography* 19:459–479.

Clogg. C. C. 1982b. "Using Association Models in Sociological Research: Some Examples." *American Journal of Sociology* 88:114–134

Clogg. C. C. 1982c. "A Note on the Identification Problem in Age–Period–Cohort Models for the Analysis of Archival Data," in O. D. Duncan & H. H. Winsborough (Eds.), *Cohorts and the Analysis of Social Change.* New York: Academic Press.

Clogg. C. C. 1982c. "Some Models for the Analysis of Association in Multi-Way Cross-Classification Having Ordered Categories." *Journal of the American Statistical Association* 77:803–815.

Clogg. C. C. 1982e. "Using Association Models in Sociological Research: Some Examples." *American Journal of Sociology* 88:114–134.

Clogg, C. C., & Eliason, S. R. 1987. "Some Common Problems in Log-Linear Analysis." *Sociological Methods and Research* 15:8–44.

Clogg, C. C., & Eliason, S. R. 1988. "A Flexible Procedure for Adjusting Rates and Proportions, Including Statistical Methods for Group Comparisons." *American Sociological Review* 53: 267–283.

Clogg. C. C.. & Eliason, S. R. 1990. "The Relationship between Labor Force Behavior and Occupational Attainment," in A. L. Kalleberg (Ed.), *Research in Stratification and Social Mobility,* Vol. 9 (pp. 159–180). Greenwich, CT: JAI.

Clogg, C. C., Eliason, S. R., & Wahl, R. 1990. "Labor Market Experiences and Labor Force Outcomes." *American Journal of Sociology* 95:1536–1576.

Clogg, C. C., Ogena, N, & Shin, H. 1991. "Labor Force Behavior in the Process of Socioeconomic Attainment: New Scales Added to Classical Models." *Social Science Research* 20:256–270.

Clogg, C. C., Petkova, E., Shihadeh, E. S. 1992. "Statistical Methods for Analyzing Collapsibility in Regression Models. *Journal of Educational Statistics* 17:31–74.

Clogg, C. C., & Sawyer, D. O. 1981. "A Comparison of Alternative Models for Analyzing the Scalability of Response Patterns," in S. Leinhardt (Ed.), *Sociological Methodology 1981* (pp. 240–280). San Francisco: Jossey–Bass.

Clogg, C. C., & Shockey, J. W. 1984. "A Note on Two Models for Mobility Tables," in R. Tomasson (Ed.), *Comparative Social Research* (pp 443–462). Greenwich, CT: JAI.

Clogg, C. C., & Shockey, J. W. 1984. "Mismatch between Occupation and Schooling: A Prevalence Measure, Recent Trends, and Demographic Analysis." *Demography* 21:235–257.

Clogg, C. C., & Shockey, J. W. 1985. "The Effect of Changing Demographic Composition on Recent Trends in Underemployment." *Demography* 22:395–414.

Clogg, C.C., Shockey, J. W., & Eliason, S. R. 1990. "A General Statistical Framework for the Adjustment of Rates." *Sociological Methods and Research* 19:156–195.

Clogg, C. C., & Sullivan, T. A. 1983. "Labor Force Composition and Underemployment Trends, 1969–1980." *Social Indicators Research* 12:117–152.

Clogg, C. C., Sullivan, T. A., & Mutchler, J. E. 1986. "On Measuring Underemployment and Inequality in the Labor Force." *Social Indicators Research* 18:375–393.

Coleman, J. S. 1990. *Foundations of Social Theory.* Cambridge, MA: Harvard University Press.

Dahrendorf, R. 1959. *Class and Class Conflict in Industrial Society.* Stanford, CA: Stanford University Press.

Duncan, B. 1979. "Change in Worker/Nonworker Ratios for Women." *Demography* 16:535–547.

Duncan, O. D. 1969. *Toward Social Reporting: Next Steps.* New York: Russell Sage Foundation.

Duncan, O. D. 1975. "Partitioning Polytomous Variables in Contingency Tables." *Social Science Research* 4:167–182.

Duncan, O. D. 1979. "How Destination Depends on Origin in the Occupational Mobility Table," *American Journal of Sociology* 84:793–803.

Duncan, O. D. 1981. "Two Faces of Panel Analysis: Parallels with Comparative Cross-Sectional and Time-Tagged Association," in S. Leinhardt (Ed.), *Sociological Methodology 1981* (pp. 65–95). San Francisco: Jossey–Bass.

Duncan, O. D., Featherman, D. L., & Duncan, B. 1972. *Socioeconomic Background and Achievement.* New York: Seminar.

Dunkerley, M. 1996. *The Jobless Economy? Computer Technology and the World of Work.* Cambridge: Polity Press.

Durand, J. D. 1948. *The Labor Force in the United States, 1890–1960.* New York: Gordon & Breach Science Publishers.

Easterlin, R. A. 1978. "What Will 1984 be Like? Socioeconomic Implications of Recent Twists in Age Structure." *Demography* 15:397–432.

Easterlin, R. A. 1980. *Birth and Fortune: The Effects of Generation Size on Personal Welfare.* New York: Basic Books.

Eliason, S. R. 1989. *CDAS: Categorica/Data Analysis System User's Manual.* Unpublished manuscript, Department of Sociology, University of Iowa.

Eliason, S. R. 1995a. "An Extension of the Sorensen–Kalleberg Theory of the Labor

Market Matching and Attainment Processes." *American Sociological Review* 60:247–271

Eliason, S. R. 1995b. "Modeling Manifest and Latent Dimensions of Association in Two-Way Cross-Classifications." *Sociological Methods and Research* 24:30–67.

England, P. 1992. *Comparable Worth: Theories and Evidence.* New York: Aldine de Gruyter.

England, P., & Farkas, G. 1986. *Households, Employment, and Gender.* New York: Aldine.

Farkas, G. 1977. "Cohort, Age, and Period Effects upon the Employment of White Females: Evidence for 1957–1968." *Demography* 14:33–42.

Farley, R. 1984. *Blacks and Whites: Narrowing the Gap?* Cambridge, MA: Harvard University Press.

Featherman, D. L., & Hauser, R. M. 1978. *Opportunity and Change.* New York: Academic Press.

Feller, W. 1968. *An Introduction to Probability Theory and its Applications* (Vol. 1, 3rd ed.). New York: Wiley.

Fienberg, S. E. 1980. *The Analysis of Cross-Classified Categorical Data* (2nd ed.). Cambridge, MA.: MIT Press.

Fienberg, S. E., &. Mason, W. M. 1978. "Identification and Estimation of Age–Period–Cohort Models in the Analysis of Discrete Archival Data," in Karl F. Schuessler (Ed.), *Sociological Methodology 1979* (pp. 1–67). San Francisco: Jossey–Bass.

Flaim. P. O. 1972. "Discouraged Workers and Changes in Unemployment." *Monthy Labor Review* 95:8–16: p 3.

Flaim, P. O. 1979. "The effect of demographic changes on the nation's unemployment rate." *Monthly Labor Review* 102:13–23.

Freeman, R. B. 1976, *The Over-Educated American.* New York: Academic Press

Gilroy, C. L. 1975. "Suppliemental Measures of Labor Force Underutilization." *Monthly Labor Review* 98:13–25.

Gilula, Z., & S. J. Haberman. 1988. "The Analysis of Multivariate Contingency Tables by Restricted Canonical and Restricted Association Models." *Journal of the American Statistical Association* 83:760–771.

Glass, D. V. (Ed.). 1954. *Social Mobility in Britain.* London: Routledge and Kegan Paul.

Glass, J., & Riley, L. 1998. "Family Responsive Policies and Employee Retention Following Childbirth." *Social Forces* 76:1401–1435.

Glenn, N. D. 1977. *Cohort Analysis.* Beverly Hills: Sage.

Goodman, L. A. 1969. "How to Ransack Social Mobility Tables and Other Kinds of Cross Classification Tables." *American Journal of Sociology* 75:1–39.

Goodman, L. A. 1971, "Partitioning of Chi-Square, Analysis of Marginal Contingency Tables and Estimation of Expected frequencies in Multidimensional Contingency Tables." *Journal of the American Statistical Association* 66:339–392

Goodman, L. A. 1972. "Some Multiplicative Models for the Analysis of Cross Classified Data," in L. LeCam, I. Neyman, & E. L. Scott (Eds.), *Proceeding of the Sixth Berkeley Symposium on Mathematical Statistics and Probability* (Vol. I, pp. 649–696). Berkeley: University of California Press.

Goodman, L. A. 1974a. "Exploratory Latent Structure Analysis Using Both Identifiable and Unidentifiable Models." *Biometrika* 61:215–231.

Goodman, L. A. 1974b. "The Analysis of Systems of Qualitative Variables When Some of the Variables Are Unobservable: Part 1: A Modified Latent Structure Approach." *American Journal of Sociology* 79:1179–1259.

Goodman, L. A. 1975. "A New Model for Scaling Response Patterns: An Application of the Quasi-Independence Concept." *Journal of the American Statistical Association* 70:755–768.

Goodman, L. A. 1978. *Analyzing Qualitative/Categorical Data*. Cambridge, MA: Abt Books.

Goodman, L. A. 1979. "Simple Models for the Analysis of Association in Cross-Classifications Having Ordered Categories." *Journal of the American Statistical Association* 74:537–552.

Goodman, L. A. 1984. *The Analysis of Cross-Classifications Having Ordered Categories*. Cambridge, MA: Harvard University Press.

Goodman, L. A. 1986. "Some Useful Extensions of the Usual Correspondence Analysis Approach and the Usual Log-Linear Models Approach in the Analysis of Contingency Tables." *International Statistical Review* 54:243–309.

Goodman, L. A. 1991. "Measures, Models, and Graphical Displays in Cross-Classified Data." *Journal of the American Statistical Association* 86:1085–1110.

Goodman L.A., & Kruskal, W. H. (1954) "Measures of association for cross-classifications." *Journal of the American Statistical Association* 49:732–764.

Gordon, D. M. 1996. *Fat and Mean: The Corporate Squeeze of Working Americans and the Myth of Managerial "Downsizing."* New York: Free Press.

Gordon, R. A. 1976. "Another Look at the Goals of Full Employment and Price Stability." in *National Commission for Manpower Policy, Demographic Trends and Full Employment* (Special Report No. 12, pp. 5–26). Washington, DC: U.S. Government Printing Office.

Haberman, S. J. 1978. *Analysis of Qualitative Data. Vol. 1. Introductory Topics*. New York: Academic Press.

Harrison, B., & Blueston, B. 1988. *The Great U-Turn: Corporate Restructuring and the Polarizing of America*. New York: Basic Books.

Hauan, S., Landale, N. S., & Leicht, K. T. 2000. "Poverty and Work Effort Among Urban Latino Men." *Work and Occupations* 27:188–222.

Hauser, P. M. 1949, "The Labor Force and Gainful Workers—Concept, Measurement, and Comparability." *American Journal of Sociology* 54:338–355.

Hauser, P. M. 1974. "The Measurement of Labour Utilization." *Malayan Economic Review* 19:1–17.

Hauser, P. M. 1977. "The Measurement of Labour Utilization—More Empirical Results." *Malayan Economic Review* 22:10–25.

Hauser, R. M., & Featherman, D. L. 1977. *The Process of Stratification: Trends and Analyses*. New York: Academic Press.

Hauser, R. M., Tsai, S., & Sewell, W. H. 1983. "A Model of Stratification with Response Error in Social and Psychological Variables." *Sociology of Education* 56:20–46.

Heckman, I. 1974. "Shadow Prices, Market Wages and Labor Supply." *Econometrica* 42:1659–1674.

Heckman, J. J. 1979. "Sample Selection Bias as a Specification Error." *Econometrica* 45:153–161.

Hodson, R., & Kaufman, R. 1982. "Economic Dualism: A Critical Review." *American Sociological Review* 47:727–789.

Holzer, H. J. 1996. *What Employers Want: Job Prospects for Less-Educated Workers.* New York: Russell Sage Foundation.

Hout, M. 1988. "More Universalism, Less Structured Mobility: The American Occupational Structure in the 1980s." *American Journal of Sociology* 93:1358–1400.

Hout, M., Brooks, C., & Manza, J. 1995. "The Democratic Class Struggle in the United States, 1948–1992." *American Sociological Review* 60: 805–828.

International Labour Office (ILO). 1957. "Measurement of Underemployment." Report IV of the Ninth International Conference of Labour Statisticians. Geneva: Author.

International Labour Office 1966. "Measurement of Underemployment: Concepts and Methods." Report IV of the Eleventh International Conference of Labor Statisticians Geneva: Author.

International Labour Office. 1976. "International Recommendations on Labour Statistics." Geneva: Author.

Jencks, D. 1992. *Rethinking Social Policy: Race Poverty and the Underclass.* Cambridge, MA: Harvard University Press.

Jencks, C., Crouse, J., & Mueser, P. 1983. "The Wisconsin Model of Status Attainment: A National Replication with Improved Measures of Ability and Aspirations." *Sociology of Education* 56:319.

Kalleberg, A. L., & Berg, I. 1987. *Work and Industry: Structures, Markets, and Processes.* New York: Plenum.

Kalleberg, A. L., & Sørensen, A. B. 1979. "The Sociology of Labor Markets." in *Annual Review of Sociology* (Vol. 5, pp. 351–379). Palo Alto, CA: Annual Reviews.

Kasarda, J. D. 1991. "National Business Cycles and Community Competition for Jobs." *Social Forces* 69:733–761.

Kaufman, Hodson, & Fligstein. 1981. "Defrocking Dualism: A New Approach to Defining Industrial Sectors." *Social Science Research* 10:1–31.

Keppel, K. G. 1981. "Mortality Differentials by Size of Place and Sex in Pennsylvania for 1960 and 1970." *Social Biology* 28:41–48.

Kenney, M., & Florida, R. 1993. *Beyond Mass Production: The Japanese System and its Transfer to the U.S.* New York: Oxford University Press.

Kerr, C., & Rostow, J. M. 1979. *Work in America: The Decade Ahead.* New York: Van Nostrand.

Keyfitz, N. 1968. *Introduction to the Mathematics of Population.* Reading, MA: Addison–Wesley.

Keyfitz, N. 1968. "Review of Measuring Underemployment, by C. C. Clogg." *American Journal of Sociology* 86:1163–1165.

Kitagawa, E. M. 1964. "Standardized Comparisons in Population Research." *Demography* 1:296–315.

Kitagawa, E. M. 1966. "Theoretical Considerations in the Selection of a Mortality Index, and Some Empirical Comparisons." *Human Biology* 38:293–308.

Kruskal, W. 1987. "Relative Importance by Averaging Over Orderings." *The American Statistician* 41:6–10.

Land, K. C. 1976. "The Role of Quality of Employment Indicators in General Social Reporting Systems," in A. D. Biderman & T. F. Drury (Eds.), *Measuring Work Quality for Social Reporting.* New York: Wiley. p. 200–234.

Lazarsfeld, P.F., & Henry, N. W. 1968. *Latent Structure Analysis.* Boston: Houghton Mifflin.

Leicht, K. T. 1998. "Work (if you can get it) and Occupations (if there are any)? What Social Scientists Can Learn from Predictions of the End of Work and Radical Workplace Change." *Work and Occupations* 25:36–48.

Leicht, K. T., Hogan, D. P., & Wendt, H. A. 2000. "Human Capital, Labor Market Decisions, and the Structure of Careers in the Post-War Era." Department of Sociology, The University of Iowa.

Leicht, K. T., Wallace, M., & Grant, D. S. 1993. "Union Prescence, Class, and Individual Earnings Inequality." *Work and Occupations* 20:429–431.

Leigh, D. 1978. *An Analysis of the Determinants of Occupational Upgrading.* New York: Academic Press.

Levitan, S. A., & Taggart, R. 1973. "Employment and Earnings Inadequacy: A Measure of Worker Welfare." *Monthly Labor Review* 96:19–27.

Levitan, S. A., & Taggart, R. 1974. *Employment and Earnings Inadequacy: A New Social Indicator.* Baltimore: Johns Hopkins University Press.

Lichter, D. T. 1988. "Racial Differences in Underemployment in American Cities." *American Journal of Sociology* 93:771–792.

Little, R. J. A., & Rubin, D. B. (1987), *Statistical Analysis with Missing Data.* Wiley New York.

Long, C. D. 1958. *The Labor Force Under Changing Income and Employment.* Princeton, NJ: Princeton University Press.

Massey, D. S. 1990. "American Apartheid: Segregation and the Making of the Underclass." *American Journal of Sociology* 96:329–357.

Massey, D. S. 1996. "The Age of Extremes: Concentrated Affluence and Poverty in the Twenty-First Century." *Demography* 33:395–413.

Mead, L. 1991. *The New Politics of Poverty.* New York: Basic Books.

Miller, S. M. 1960. "Comparative Social Mobility." *Current Sociology* 9:1–89.

Mincer, I. 1962. "Labor Force Participation of Married Women," in *Aspects of Labor Economics: A Conference of the Universities National Bureau Committee for Economic Research* (pp. 63–97). Princeton, NJ: Princeton University Press.

Moore, T. P. 1996. *The Disposable Workforce: Worker Displacement and Employment Instability in America.* New York: Aldine De Gruyter.

Myrdal, G. 1968. *Asian Drama.* New York: Pantheon.

National Bureau of Economic Research. 1957. *The Measurement and Behavior of Unemployment.* Princeton, NJ: Princeton University Press.

National Center for Health Statistics. 1979. *Vital Statistics of the United States: 1975.* Vol. II. Mortality. Part A. Washington, DC: U.S. Government Printing Office.

National Commission on Employment and Unemployment Statistics. 1979. *Counting the Labor Force.* Washington, DC: U.S. Government Printing Office.

New York Times. 1996. *The Downsizing of America.* New York: Times Books

Otto, L. B., Speener, K. I., & Call, V. R. A. 1980a. *Career Line Prototypes* (1st ed.). Omaha, NB: Boys' Town Center for the Study of Youth Development.

Otto, L. B., Speener, K. I., & Call, V. R. A. 1980b. *Career Line Prototypes* (2nd ed.). Omaha, NB: Boys' Town Center for the Study of Youth Development.

Perry, C. 1970. "Changing Labor Markets and Inflation." *Brookings Papers on Economic Activity* 3:411–441.

Perrucci, C., Perrucci, R., Targ, D., & Targ, H. 1988. *Plant Closings: International Context and Social Costs.* New York: A. de Gruyter.

Petersen, T., & Morgan, L. A. 1995. "Separate and Unequal: Occupation-Establishment Sex Segregation and the Gender Wage Gap." *American Journal of Sociology* 101:329–365.

Phelps, E. 1997. *Rewarding Work: How to Restore Participation and Self-Support to Free Enterprise.* Cambridge, MA: Harvard University Press.

Pub. L. No. 94- 144: 1977, 90 Stat. 1483, 29 U.S.C. 952 (Suppl., Note).

Pullum, T. 1975. *Measuring Occupational Inheritance.* New York: Elsevier.

Pullum, T. W. 1977. "Parametrizing Age, Period and Cohort Effects: An Application to U.S. Delinquency Rates," in Karl F. Schuessler (Ed.), *Sociological Methodology 1978* (pp. 116–140). San Francisco: Jossey–Bass.

Pullum, T. W. 1980. "Separating Age, Period and Cohort Effects in White U.S. Fertility, 1920–1970." *Social Science Research* 9:225–244.

Quinn, R. P., & Staines, G. L. 1979. "The 1977 Quality of Employment Survey." Institute for Social Research, Survey Research Center, Ann Arbor, MI.

Reskin, B. F., & Roos, P. A. 1990. *Job Queues, Gender Queues: Explaining Women's Inroads into Male Occupations.* Philadelphia: Temple University Press.

Ridgeway, C. 1997. "Interaction and the Conservation of Gender Inequality: Considering Employment." *American Sociological Review* 62:218–235.

Rifkin, J. 1995. *The End of Work.* New York: Tarcher/Putnam.

Robinson, J. 1936. "Disguised Unemployment." *Economic Journal* 46:235–237.

Rosenfeld, R. A., & Kalleberg, A. L. 1991. "Gender Inequality in the Labor Market." *Acta Sociologica* 34:207–325.

Ryder. N. B. 1964. "The Process of Demographic Translation." *Demography* 1:74–82.

Ryder. N. B. 1965. "The Cohort as a Concept in the Study of Social Change." *American Sociological Review* 30:843–861.

Ryder. N. B. 1969. "The Emergence of a Modern Fertility Pattern: United States 1917–66," in S. J. Behrman, L. Corsa, Jr., & R. Freeman (Eds.), *Fertility and Family Planning* (pp. 99–126). Ann Arbor: University of Michigan Press.

Schervish, P. G. 1983. *Vulnerability and Power in Market Relations: The Structural Determinants of Unemployment.* New York: Academic Press.

Schor, J. 1992. *The Overworked American.* New York: Basic Books.

Schor, J. 1998. *The Overspent American.* New York: Basic Books.

Seashore, S. E., & Taber, T. D. 1976. "Job Satisfaction Indicators and Their Correlates," in A. D. Biderman & T. G. Drury (Eds.), *Measuring Work Quality for Social Reporting.* New York: Wiley.

Smith, R. E. 1977. "A Simulation Model of the Demographic Composition of Employment, Unemployment and Labor Force Participation," in R. E. Ehrenberg (Ed.), *Research in Labor Economics* (pp. 259–304). Greenwich, CT: JAI Press.

Smith, R. E., Vanksi,J., & Hott, C. 1974. "Recession and the employment of demographic groups." *Brookings Papers on Economic Activity* 2:727–760.

Sørensen, A. B. 1983. "Sociological Research on the Labor Market: Conceptual and Methodological Issues." *Work and Occupations* 10:261–287.

Spenner, K. I., Otto, L. B., & Call, V. A. R. 1982. *Career Lines and Careers.* Lexington, MA: D. C. Heath.

Spilerman, S. 1977. "Careers, Labor Market Structure, and Socio-Economic Achievement." *American Journal of Sociology* 85:551–593.

Stevens, G., & Hoisington, E. 1987. "Occupational Prestige and the 1980 U.S. Labor Force." *Social Science Research* 16:74–105.

Stolzenberg, R. 1994. "Educational Continuation by College Graduates." American Journal of Sociology 99:1042–1077.

Stolzenberg, R. M., & Relles, D. A. 1990. "Theory Testing in a World of Constrained Research Design: The Significance of Heckman's Censored Sampling Bias Correction for Nonexperimental Research." *Sociological Methods and Research* 19:1–29.

Sullivan, T. A. 1978. *Marginal Workers, Marginal Jobs: The Underutilization of American Workers*. Austin: University of Texas Press.

Sullivan, T. A., & Hauser, P. M. 1979. "The Labor Utilization Framework: Assumptions, Data, and Policy Implications," in *National Commission on Employment and Unemployment Statistics. Counting the Labor Force*, Appendix Vol. 1, Concepts and Data Needs (pp. 246–271). Washington, DC: U.S. Government Printing Office.

Svalastoga, K. 1959. *Prestige, Class, and Mobility*. London: Heineman.

Sweet, I. A. 1973. *Women in the Labor Force*. New York: Seminar Press.

Tam, T. 1997. "Sex Segregation and Occupational Gender Inequality in the United States: Devaluation or Specialized Training." *American Journal of Sociology* 102:1652–1692.

Taueber, K.E. 1976. "Demographic Trends Affecting the Future Labor Force," in *National Commission for Manpower Policy. Demographic Trends and Full Employment*. Special Report No. 12 (pp. 101–191). Washington, DC: U.S. Government Printing Office.

Tomaskovic-Devey, D. 1993. *Gender and Racial Inequality at Work*. Ithaca, NY: ILR.

Turnham, D. 1971. *The Employment Problem in Less Developed Countries*. Paris: Development Centre of the Organization for Economic Co-Operation and Development.

U.S. Bureau of Census. 1980b "The Social and Economic Status of the Black Population in the United States, 1790–1978." Current Population Reports, Series P-25, No. 80. Washington, DC: U.S. Government Printing Office.

U.S. Bureau of Commerce. 1964. The Current Population Survey—A Report on Methodology. Washington, DC: U.S. Government Printing Office.

U.S. Bureau of Commerce. 1967. Concepts and Methods Used in Manpower Statistics from the Current Population Survey. Washington, DC: U.S. Government Printing Office.

U.S. Bureau of Commerce. 1969. On the Evolution of Manpower Statistics. Kalamazoo, MI: The Upjohn Institute.

U.S. Bureau of Commerce. 1975. Characteristics of the Low-Income Population: 1973. Washington, DC: U.S. Government Printing Office.

U.S. Bureau of Commerce. 1978. The Current Population Survey—Design and Methodology. Washington, DC: U.S. Government Printing Office.

U.S. President's Committee to Appraise Employment and Unemployment Statistics. 1962. Measuring Employment and Underemployment. Washington, DC: U.S. Government Printing Office.

Vedder, R. K., & Gallaway, L. 1992. "Racial Differences in Unemployment in the United States, 1890–1990." *The Journal of Economic History* 52:696–702.

Vietorisz, T., Mier, R., & Giblin, J. E. 1975. "Subemployment: Exclusion and Inadequacy Indexes." *Monthly Labor Review* 98:3–12.

Wachter, M. L. 1976. "The Demographic Impact on Unemployment: Past Experience and the Outlook for the Future," in National Commission for Manpower Policy. Demographic Trends and Full Employment. Special Report No. 12 (pp. 27–98). Washington, DC: U.S. Government Printing Office.

Wallace, M., & Rothschild. 1988. *Deindustrialization and the Restructuring of American Industry.* Greenwich, CT: JAI Press.

Weiss, Y., & Lillard, L. A. 1978. "Experience, Vintage, and Time Effects in the Growth of Earnings: American Scientists." *Journal of Political Economy* 86:427–447.

Welch, F. 1979. "Effects of Cohort Size on Earnings." *Journal of Political Economy* 87:565–598.

Wernick, M. S., & Mcintire, J. L. 1980. "Employment and Labor Force Growth: Recent Trends and Future Prospects," in Joint Economic Committee. Congress of the United States, Special Studies on Economic Change. Vol. 1. Human Resources and Demographics: Characteristics of People and Policy (pp. 101–152). Washington. DC: U.S. Government Printing Office.

Wilson, F. D., Tienda, M., & Wu, L. 1995. "Race and Unemployment: Labor Market Experiences of Black and White Men, 1968–1988." *Work and Occupations* 22:245–270.

Wilson, W. J. 1987. *The Truly Disadvantaged: The Inner City, the Underclass, and Public Policy.* Chicago: University of Chicago Press.

Wilson, W.J. 1996. *When Work Disappears: The World of the Urban Poor.* New York: Knopf.

Winsborough, H. H. 1978. "Statistical Histories of the Life Dycle of Birth Dohorts: The Transition from Schoolboy to Adult Male," in R. E. Taeuber, L. L. Bumpass, & J. A. Sweet (Eds.), *Social Demography* (pp. 213–245). New York: Academic Press.

Wunsch, G. J., & Termote, M. G. 1978. *Introduction to Demographic Analysis: Principles and Methods.* New York: Plenum Press.

Index

263

Military personnel, 64
Minorities and immigrants
  minorities in the labor force, 9, 45, 243–244
Mobility tables, 96
  Danish mobility tables, 97, 99
  British mobility tables, 97, 99
Multiway contingency tables, 142

National Bureau of Economic Research, 32
National Longitudinal Survey of Youth, 244
National employment policies, 5
New entrants to the labor force, 34, 37, 194–195, 222–223, 236–237
Nonmarginal latent classes, 8, 9
Nonworkers, 224, 237–238
Normative social indicators of underemployment, 56

Occupational careers, 164
Occupational mismatch, 6, 7, 8, 10, 20, 24, 29, 31, 39–40, 47, 56, 73, 77, 84–86, 172
OCG-II sample, 12, 158–163
On-the-job training, 21

Panel Study of Income Dynamics, 244
Part-time employment, 20, 174, 189–192, 220–221, 224, 227, 232–233
Part-year employment, 35
Partial regression coefficients, 112
Period effects, 57, 62–65, 66, 70, 80, 87
Policy debates
  about equalizing labor market experiences, 199
  about low-income workers, 25
  Tax incentives to produce full-time jobs, 239
  "welfare-to-work," 240
Policy proposals to reduce underemployment, 48
  constraints versus choices, 248
  equalizing labor market experiences, 249
  freedom of choice, 248
  improving labor market matching, 250
  overuse of full-time workers, 249
  reducing inequality, 248
  reducing working poverty, 250
Post-industrial workforce, 244
Poverty research, 141

Poverty status of households, 21, 38
Productivity of work, 16, 19
  ILO definition of, 31, 33

Quasi-latent structure models, 12
Quasi-perfect mobility models, 104, 116

Racial and gender inequality in labor force outcomes, 167–215
Racial stereotypes, 198
Rate adjustment techniques in log-linear models, 92–94
RC association models, 128–131,
  conditional independence, 179
  constraints imposed on, 202–203
  disaggregating by sex and age, 147–148
  estimation of, 200–203
  geometric representation of, 131–133, 143
  group-specific components in, 178
  influence measures calculated from, 179–181
  interpreting score parameters, 130–131
  intrinsic dimensions revealed by, 131, 146, 202
  model estimation, 146–147
  occupational heterogeneity in, 225–226, 238
  parameter values from, 129–130, 142, 151
  partitioning likelihood ratio statistics from, 204–205
  purging methods used with, 169, 184, 227
  similarity to log-linear models, 201
  time-specific components in, 178, 225–226, 238
Recessions, 123
Reentrants to the labor force, 34, 37
Research agenda for studying the labor force,
  access to skilled jobs with fringe benefits, 246
  credential inflation and educational mismatch, 246–247
  flexibility and debureaucratization of the workplace, 247
  monitoring change across occupational groups, 245
  studying labor force careers, 244
Reservation wages, 87
Reserve labor, 122

PLENUM STUDIES IN WORK AND INDUSTRY
COMPLETE CHRONOLOGICAL LISTING

Series Editors:
Ivar Berg, *University of Pennsylvania, Philadelphia, Pennsylvania*
and Arne L. Kalleberg, *University of North Carolina, Chapel Hill, North Carolina*

WORK AND INDUSTRY
Structures, Markets, and Processes
Arne L. Kalleberg and Ivar Berg

WORKERS, MANAGERS, AND TECHNOLOGICAL CHANGE
Emerging Patterns of Labor Relations
Edited by Daniel B. Cornfield

INDUSTRIES, FIRMS, AND JOBS
Sociological and Economic Approaches
Edited by George Farkas and Paula England

MATERNAL EMPLOYMENT AND CHILDREN'S DEVELOPMENT
Longitudinal Research
Edited by Adele Eskeles Gottfried and Allen W. Gottfried

ENSURING MINORITY SUCCESS IN CORPORATE MANAGEMENT
Edited by Donna E. Thompson and Nancy DiTomaso

THE STATE AND THE LABOR MARKET
Edited by Samuel Rosenberg

THE BUREAUCRATIC LABOR MARKET
The Case of the Federal Civil Service
Thomas A. DiPrete

ENRICHING BUSINESS ETHICS
Edited by Clarence C. Walton

LIFE AND DEATH AT WORK
Industrial Accidents as a Case of Socially Produced Error
Tom Dwyer

WHEN STRIKES MAKE SENSE—AND WHY
Lessons from Third Republic French Coal Miners
Samuel Cohn

THE EMPLOYMENT RELATIONSHIP
Causes and Consequences of Modern Personnel Administration
William P. Bridges and Wayne J. Villemez

LABOR AND POLITICS IN THE U.S. POSTAL SERVICE
Vern K. Baxter

NEGRO BUSINESS AND BUSINESS EDUCATION
Their Present and Prospective Development
Joseph A. Pierce
Introduction by John Sibley Butler

SEGMENTED LABOR, FRACTURED POLITICS
Labor Politics in American Life
William Form

PLENUM STUDIES IN WORK AND INDUSTRY
COMPLETE CHRONOLOGICAL LISTING